Quantum Chromodynamic

Edited by Zbigniew Piotr Szadkowski

Published in London, United Kingdom

IntechOpen

Supporting open minds since 2005

Quantum Chromodynamic
http://dx.doi.org/10.5772/intechopen.90972
Edited by Zbigniew Piotr Szadkowski

Contributors
Zbigniew Piotr Szadkowski, Zalak Shah, Ajay Kumar Rai, Nancy Lynn Bowen, Mikael Souto Maior De Sousa, Rômulo Rodrigues Da Silva, Richard Oldani

First published in London, United Kingdom, 2021 by IntechOpen
IntechOpen is the global imprint of INTECHOPEN LIMITED, registered in England and Wales, registration number: 11086078, 5 Princes Gate Court, London, SW7 2QJ, United Kingdom
Printed in Croatia

British Library Cataloguing-in-Publication Data
A catalogue record for this book is available from the British Library

Additional hard and PDF copies can be obtained from orders@intechopen.com

Quantum Chromodynamic
Edited by Zbigniew Piotr Szadkowski
p. cm.
Print ISBN 978-1-83968-312-1
Online ISBN 978-1-83968-313-8
eBook (PDF) ISBN 978-1-83968-314-5

Meet the editor

Zbigniew Szadkowski obtained his Ph.D. in Theoretical Physics from the University of Lodz, Poland, in 1987 with his thesis, “Quark mixing in the SU(4) x SU(4) and SU(6) x SU(6) chiral symmetries.” In 2007, he obtained a DSci in Experimental Physics based on the thesis, “Triggers in the Pierre Auger Observatory.” Since 1998, Dr. Szadkowski has been working in the Pierre Auger Observatory designing the 2nd-level trigger implemented in all 24 fluorescence telescopes, the 1st-level trigger implemented in all 1680 surface detectors, the Front-End Boards with ACEX (800 pieces) and Cyclone (900 pieces) FPGAs, and many algorithms for triggers. He worked at the Michigan Technological University, USA from 1999 to 2002, the College de France, Paris from 2002 to 2003, and the Bergische University of Wuppertal, Germany from 2003 to 2006.

Contents

Preface

Quantum chromodynamics (QCD) is a quantum field theory that describes strong interactions between quarks and gluons. It is the SU(3) gauge theory of the current Standard Model for elementary particles and forces.

This book contains a set of essays on various aspects of QCD, including the various natures of quarks from both theoretical and experimental points of view. The authors believe that ingenious and effective use of the newest achievements in QCD is key to the development of not only physics but also science in general.

This book promotes cooperation between various disciplines as well as within larger research communities. The editors believe that this book will interest both avid experimentalists and theoreticians. We believe that quantum chromodynamics is an integrated discipline of the entire physics, an entry to Grand Unified Theories which ultimately allow integrating all fundamental interactions into a single, brilliant theory describing the entire world.

Zbigniew Piotr Szadkowski
University of Łódź,
Łódź, Poland

Section 1

Introduction

Chapter 1

Introductory Chapter: Quantum Chromodynamic

Zbigniew Piotr Szadkowski

1. Hypothesis of quarks

In the 1950s, the list of elementary particles was very long. Apart from "old" particles such as protons, neutrons, and electrons, the list contained also muons, mesons $\pi^{\pm}$, π^{0}, "heavy" mesons K, hyperon Λ, $\Sigma^{\pm}$, Σ^{0}, Ξ^{0}, Ξ^{-}. These particles decay in weak interactions. Their average lifetime was at a level of 10^{-10} s. However, some other very short-time particles were discovered (called resonances) decaying in strong interactions with the lifetime at a level of 10^{-23} s.

Among many proposals introducing some order in the elementary particle world, the most significant was the idea of Murray Gell-Mann. He proposed to introduce a new quantum number—"strangeness." This name appeared too frivolous for the Physical Review editor who demanded a more appropriate description. Finally, in the printed version of the paper, the term "unstable" was used. Gell-Mann was so disgusted that he did not publish more in Physical Review.

The English theoretician Richard Dalitz mentioned that the physics of mesons π was the most important for physicists at that time, while strange particles were treated as some type of squalidity not important in nucleus interactions. When Dalitz published a very good paper on some aspects of decay of some meson, he was warned that the development of such suspicious "theories" can be fatal for his carrier.

In 1961, Gell-Mann [1] and independently Israeli physicist Yuval Ne'eman [2] developed a classification of particles based on the SU(3) group. This symmetry is an extension of isospin symmetry introduced in 1932 by Heisenberg who noticed that proton and neutron can be treated as two different states of the same particle: nucleon. They differ only by the electric charge. In the world without the electromagnetic interactions, proton and neutron were indistinguishable.

In the SU(3) symmetry differs between themselves not only by electric charge but also by the strangeness. This formalism allowed Gell-Mann to predict the existence of the hyperon with a triple strangeness denoted as Ω^{-}. In 1964, this particle was discovered and its features were exactly as predicted by SU(3) formalism.

In 1964, Gell-Mann introduced a very brave hypothesis that all hadrons are built from sub-elementary components called "quarks" [3]. Simultaneously, a similar idea was proposed Georg Zweig. But his idea was published in the internal CERN bulletin and did not get a wide popularity. Gell-Mann assumed that three types of quarks exist in nature: "up," "down," and "strange." They would be fermions with the baryon number ⅓. The electric chargé of u-quark was + ⅔, and both d and s quarks were electric chargé –⅓. Simultaneously with quarks, anti-quarks should exist. Hadrons would consist of 3 quarks, mesons from a pair quark-antiquark. A majority of physicists in that time considered Gell-Mann idea as ridiculous and nonsensical. A society's prejudice was so strong that scientific promotions were blocked for scientists developing such nonsenses. Nevertheless, the quark model slowly became step by step more successful. It explains in a very simple way that the cross section between nucleon-nucleon interactions is 1.5 times bigger than the cross section between pions and nucleons. It was fast noticed that the introduction of the new quantum number is needed to remain the Pauli exclusion principle. For example, according to the Gell-Mann scheme Hyperon Ω^- should be built from three "s" quarks in the same state. Oscar Greenberg [4], Moo-Young Han, and Yoichiro Nambu [5] introduced in 1964 an idea of color charge (shortly a color). This idea removed difficulties with the Pauli principle.

From the time of introduction of quarks as hypothetic components of hadrons, the important problem was for an explanation whether hadrons crashing for very high energies can free quarks and we can see them as free particles. A matter ionization is proportional to the electric charge. Traces in cloud chambers remaining by free quarks should correspond to 1/9 and 4/9 in comparison with free electrons. Many research centers have looking for such traces and even some centers reported these investigations as successful; however, all appeared false. When any attempts of free quarks finding failed and theoretical assumptions were considered as doubtful, quarks started to be treated as fiction and a singularity. Even the author of quarks Murray Gell-Mann certified in 1972 that "it is possible to build the hadron theory based on quarks as fiction objects."

The best tools for studying the nucleon structure are point particles, for example, electrons. In 1933, we knew that nucleons have some structure because they have much bigger magnetic moment than calculated from Dirac equations for the point particle.

In 1953, Robert Hofstadter marked out a density distribution of the electric charge for many nucleuses and in 1954, also for proton and neutron giving their "electric radius" (squared averaged radius their charge density distribution ~0.74×10^{-15} m). Results showed that nucleons have some continuous structure.

In 1967, Jerome Friedman, Henry Kendall, and Richard Taylor started experiments at Stanford with a deep inelastic scattering of electrons on protons. Results were difficult to understand. Theoretical speculations suggested that some point structures can exist inside nucleons. At the same time, Richard Feynman and James Bjorken try to explain the Stanford results. Feynman developed the parton model, which has been confirmed in the next experiments and is used up today. Parton can be identified with quarks.

In 1974, the so-called "November revolution" significantly concurred to acceptance of the quark model. Two experimental teams from Stanford (conducted by Burton Richter) and Brookhaven (conducted by Samuel Ting) published simultaneously [6, 7] on a discovery of a new particle with a mass three times bigger than a proton. Teams used quite different methods and did not know each other on the "concurrent" experiment. The features of a newly discovered particle can be explained only by an assumption that it consists of a quark of a new type. The existing of the fourth quark was considered earlier, now this was an irrefutable proof. Symbols proposed for a new particle were different and neither Stanford nor Brookhaven teams were convinced to use a unified symbol. The particle is known as J/ψ. The fourth quark was called as "charmed."

Soon, proofs on the existence of new two quarks were obtained. Leon Liederman's team from Fermilab discover the fifth quark (denoted as "beauty" or "bottom"). In 1995, the sixth quark (denoted as "true" or "top") was discovered. At present, we accept the existence of three quark doublets: (u,d), (c,s), and (b,t). We are almost sure that the third family is the last and the fourth generation of quarks does not exist.

The only experimental results were not enough for the physics society. A theory explaining quarks interactions was expected.

2. The road toward the Standard model

In 1971, Gerardus t'Hooft [8, 9] proved that the gauge theories (introduced in 1954 by Chen-Ning Yang and Robert Mills [10]) can be renormalized as quantum electrodynamic. In 1973, David Gross and Frank Wilczek [11] from Princeton and independently David Politzer [12] from Harvard discovered quark asymptotic freedom. An interaction force between quarks increases with a distance between them and vanishes when quarks approach each other. Quarks on very small distances can be treated as almost free. The theory explained the problem of quark confinement. Quarks, as a component of hadrons, must be confined inside hadrons.

Several months later, Murray Gell-Mann, Harald Fritsch, and Heinrich Leutwyler [13] introduced carriers of quark interactions—the octet with the colored charge. This was the beginning of Quantum Chromodynamics—QCD [14].

One of the very important parts of the QCD is the theory of electro-weak interactions developed by Sheldon Glashow [15], Abdus Salam [16], and Steven Weinberg [17] known as the Standard model. This theory unified the electromagnetic interactions carrying by photons with the weak interactions carrying by intermediate bosons $W^{\pm}$ and Z^0. The last bosons were considered as an intermediary in neutral currents. In 1973, the expected neutral currents and in 1983 charged bosons W^+ and W^- and neutral Z^0 were discovered.

In QCD virtual gluons "anti-screen" color charge in a vacuum and this effect dominates on screening of electric charge by quarks. It means that the color charge, which in a big distance is large, has the source in a weak charge on small distances and aims to zero where distances between quarks also aim to zero. The force between quarks increases together with a distance between them.

The Standard Model agrees with experiments with a high precision. However, it is not a final version because it contains some free parameters that have to be taken from experimental results. Nevertheless, it is a fantastic tool describing the micro-world.

Missing elements of the Standard Model is a mechanism of mass generation by particles. Peter Higgs [18] in 1964 proposed a mechanism assuming the existence of super-heavy particle (H boson). On July 4, 2012 two experiments ATLAS and CMS announced a discovery of the Higgs boson in LHC experiments in CERN. The mass is 125.3 ± 0.4 GeV/c^2. It is a massive scalar boson with zero spin, no electric charge, and no color charge. It is also very unstable, decaying into other particles almost immediately.

In 1974, Howard Georgi and Shelton Glashow [19] proposed a first Grand Unified Theory (GUT) based on the simple Lie group SU(5). The motivation of GUT is the fact that the electric charge of electrons and protons is the same with extremely high precision but this feature is not explained by the Standard Model.

The strong and weak interactions in the Standard Model are based on gauge SU(3) and SU(2) symmetries, respectively. The weak hypercharge interaction is described by an Abelian symmetry U(1). The strong and weak interactions might be unified in one Grand Unified interaction described by a single, larger simple symmetry group containing also the Standard Model. This would automatically predict the quantized nature and values of all elementary particle charges.

The simplest group containing SU(3) × SU(2) as a candidate for the GUT is SU(5). The GUT symmetries allow a reinterpretation of known particles, like the photon, W and Z bosons, and gluon, as different states of a single particle field. The next simple Lie group that contains the standard model is SO(10). However, there are several difficulties in comparing theories with the experimental data. Model scenarios for sources of Ultra High-Energy Cosmic Rays (UHECRs), in which the observed particles are produced by the decay of other particles (top-down models), lead to large secondary fluxes of photons and neutrinos. In contrast, models in which the production of photons and neutrinos originates from secondaries generated by the propagation in the cosmic background (GZK effect) lead to much lower fluxes. The current flux limits rule out, or strongly disfavor, that top-down models can account for a significant part of the observed UHECR flux.

Author details

Zbigniew Piotr Szadkowski
Faculty of Physics and Applied Informatics, University of Łódź, Łódź, Poland

*Address all correspondence to: zbigniew.szadkowski@fis.uni.lodz.pl

References

[1] Gell-Mann M. The Eightfold Way: A Theory of Strong Interaction Symmetry (Report). Pasadena, CA: California Inst. of Tech., Synchrotron Laboratory; 1961

[2] Ne'eman Y. Derivation of strong interactions from a gauge invariance. Nuclear Physics. 1961;**26**(2):222-229

[3] Gell-Mann M. A schematic model od baryons and mesons. Physics Letters. 1964;**8**:214

[4] Greenberg OW. Spin and unitary-spin independence in a Paraquark model of baryons and mesons. Physical Review Letters. 1964;**13**:598

[5] Han MY, Nambu Y. Three-triplet model with double SU(3) symmetry. Physics Review. 1965;**139**:B1006

[6] Aubert JJ et al. Experimental observation of heavy particle. Physical Review Letters. 1974;**33**:1404

[7] Augustin JE et al. Discovery of a narrow resonance in $e^+ e^-$ annihilation. Physical Review Letters. 1974;**33**:1406

[8] 't Hooft G. Renormalization of massless Yang-Mills fields. Nuclear Physics B. 1971;**33**(1):173

[9] 't Hooft G. Renormalizable Lagrangians for massive Yang-Mills fields. Nuclear Physics B. 1971;**35**(1):167

[10] Yang CN, Mills R. Conservation of isotopic spin and isotopic gauge invariance. Physical Review. 1954;**96**(1):191

[11] Gross DJ, Wilczek F. Ultraviolet behaviour of non-abelian gauge theory. Physical Review Letters. 1973;**30**:1343

[12] Politzer HD. Reliable perturbative results for strong interactions. Physical Review Letters. 1973;**30**:1346

[13] Fritzsch H, Gell-Mann M, Leutwyler H. Advantages of the color octet gluon pictures. Physics Review. 1973;**B47**:365

[14] Weinberg S. Non-abelian gauge theories for the strong interactions. Physical Review Letters. 1973;**32**:494

[15] Glashow SL. Partial-symmetries of weak interactions. Nuclear Physics. 1961;**22**(4):579

[16] Salam A, Svartholm N, editors. Elementary Particle Physics: Relativistic Groups and Analyticity. Eighth Nobel Symposium. Stockholm: Almquvist and Wiksell; 1968. p. 367

[17] Weinberg S. A model of leptons. Physical Review Letters. 1967;**19**(21):1264

[18] Higgs PW. Broken symmetries and the masses of gauge bosons. Physical Review Letters. 1964;**13**(16)

[19] Georgi H, Glashow SL. Unity of all elementary particle forces. Physical Review Letters. 1974;**32**(8):438-441

Section 2

Quark Topics

Chapter 2

Quarks Mixing in Chiral Symmetries

Zbigniew Piotr Szadkowski

Abstract

We discuss a subject of the quarks mixing in $SU_4 * SU_4$ and $SU_6 * SU_6$ symmetries trying to calculate the quarks mixing angles and the complex phase responsible for the CP non-conservation on the basis of the Gell-Mann Oakes Renner model. Assuming symmetry breaking in a limit of exact sub-symmetries for simultaneous quarks rotations in both electric charge sub-spaces we can estimate all mention above parameters. A perfect agreement of the experimental value of the Cabibbo angles with a sum of simultaneous quarks mixing angles in doublets (u,c) and (d,s) in the $SU_4 * SU_4$ symmetry suggests that a quarks mixing is realized in a maximal allowed range. The same assumption used for the $SU_6 * SU_6$ and a simultaneous maximal allowed quarks mixing in both electric charge triplets (u,c,t) and (d,s,b) gives a perfect agreement with the experimental value of the Cabibbo angle and estimation on the angles Θ_2 and Θ_3 as well as a bond for the complex phase δ.

Keywords: quarks mixing, chiral symmetries, Cabibbo angle, Kobayashi-Maskawa mixing matrix, symmetry breaking

1. Introduction

1.1 Quarks mixing in chiral $SU_n * SU_n$ broken symmetry in the limit of exact $SU_k * SU_k$ symmetry

The hierarchy of chiral symmetry breaking [1–3] has been investigated since seventies of the previous century [4–8]. The symmetry breaking and mixing of quarks are connected with the rotation of quark currents and Hamiltonian densities. The determination of the rotation angle becomes an important problem. For the first time the procedure of chiral symmetry breaking, based on the Gell-Mann, Oakes, Renner (GMOR) model [9] has been used in $SU_3 * SU_3$ symmetry in the limit of exact $SU_2 * SU_2$ symmetry [4] to determine the value of the Cabibbo angle [10]. The transformation of rotation is connected with the seventh generator of the SU_3 group. After the charmed particles have been discovered the $SU_3 * SU_3$ symmetry is no longer adequate to describe the strong interactions. The $SU_4 * SU_4$ symmetry introduced earlier [11] to explain the behavior of charged and neutral currents becomes quite satisfactory model describing the hadron world. The problem of determining the Cabibbo angle in $SU_4 * SU_4$ symmetry has arisen. It is considered in [5–6] and the method of calculating the Cabibbo angle in $SU_4 * SU_4$ symmetry is described in [7]. It is known that the formula describing the rotation angle is not changed if the symmetry is extended. This is not unexpected because the Cabibbo angle is connected with the mixing of the d and s quarks and the rotation is performed around the seventh axis in SU_3 subspace too.

The problem of chiral $SU_4 * SU_4$ symmetry breaking in the limit of exact $SU_2 * SU_2$ symmetry is considered in [6]. Symmetry breaking is connected with the transformation of rotation around the tenth axis in SU_4 space. The rotation angle is determined in [7].

The other variant of the $SU_4 * SU_4$ symmetry breaking in the limit of the exact $SU_3 * SU_3$ symmetry is described in [8]. It is connected with the rotation around the fourteenth axis in SU_4 space. In this paper we introduce the general method of rotation angle description in the broken $SU_n * SU_n$ symmetry. The chiral $SU_n * SU_n$ symmetry is broken according to the GMOR model. In the first step we introduce the Hamiltonian density breaking $SU_n * SU_n$ symmetry but invariant under $SU_k * SU_k$ symmetry. In the second step we introduce quark mixing and the resulting exact symmetry is $SU_{k-1} * SU_{k-1}$ The particular investigation of cases like the above is not necessary.

The generalized GMOR model is used. It is assumed that by enlargement to a higher symmetry the new quantum numbers are the charges (as for example: electric charge, strangeness, charm but not isospin). Then the $SU_n * SU_n$ symmetry breaking Hamiltonian density can be written as a linear combination of diagonal operators u^i.

$$H_E = \sum_{j=1}^{n} c_{j^2-1} u^{j^2-1} \tag{1}$$

where the scalar densities $u^i = \overline{q}\lambda^i q$ and pseudo-scalar densities $v^i = i\overline{q}\lambda^i \gamma_5 q$ satisfy the equal-time commutation rules

$$\begin{aligned} \left[Q^i, u^j\right] &= if_{ijk} u^k \qquad & \left[Q^i, v^j\right] &= if_{ijk} v^k \\ \left[\overline{Q}^i, u^j\right] &= id_{ijk} v^k \qquad & \left[\overline{Q}^i, v^j\right] &= -id_{ijk} u^k \end{aligned} \tag{2}$$

where f_{ijk} are the structure constants, d_{ijk} - symmetric generators of the $SU_n * SU_n$ group. If the $SU_k * SU_k$ symmetry is exact then

$$\partial^\mu V^i_\mu = \partial^\mu A^i_\mu = 0 \qquad (i = 1, 2, \dots, k^2 - 1) \tag{3}$$

In the GMOR model the divergences of currents can be calculated as follows

$$\partial^\mu V^i_\mu = i\left[H_E, Q^i\right] \qquad \partial^\mu A^i_\mu = i\left[H_E, \overline{Q}^i\right] \tag{4}$$

We require that the $SU_k * SU_k$ symmetry be exact, then the following constraints are obeyed

$$\begin{aligned} & c_{j^2-1} = 0 \qquad (j = 2, \dots, k) \\ & \sqrt{\frac{2}{n}} c_0 + \sum_{j=k+1}^{n} \sqrt{\frac{2}{j(j+1)}} c_{j^2-1} = 0 \end{aligned} \tag{5}$$

The symmetry breaking Hamiltonian density can be written as follows

$$H_E = c_0\left(u^0 - \sqrt{n-1} u^{n^2-1}\right) + \sum_{j=k+1}^{n-1} c_{j^2-1}\left(u^{j^2-1} - \sqrt{\frac{n(n-1)}{j(j-1)}} u^{n^2-1}\right) \tag{6}$$

Using the standard representation of λ matrices one obtains

$$u^0 = \sqrt{\frac{2}{n}} \sum_{j=1}^{n} \overline{q}_j q_j \tag{7}$$

$$u^{j^2-1} = \sqrt{\frac{2}{j(j-1)}} \left(\sum_{l=1}^{j-1} \overline{q}_l q_l - (j-1) \overline{q}_j q_j \right) \tag{8}$$

$$u^0 - \sqrt{n-1}\ u^{n^2-1} = \sqrt{2n}\ \overline{q}_n q_n \tag{9}$$

$$u^{j^2-1} - \sqrt{\frac{n(n-1)}{j(j-1)}} u^{n^2-1} = \sqrt{\frac{2}{j(j-1)}} \left((n-1)\ \overline{q}_n q_n - j\ \overline{q}_j q_j - \sum_{l=j+1}^{n-1} \overline{q}_l q_l \right) \tag{10}$$

Let us note that the term $\overline{q}_k q_k$ does not exist in Eq. 11.

$$\begin{aligned} H_E = & \left(\sqrt{2n} c_0 + (n-1) \sum_{j=k+1}^{n-1} c_{j^2-1} \sqrt{\frac{2}{j(j-1)}} \overline{q}_n q_n \right) \\ & - \sum_{j=k+1}^{n-1} c_{j^2-1} \sqrt{\frac{2}{j(j-1)}} \left(j\ \overline{q}_j q_j + \sum_{l=j+1}^{n-1} \overline{q}_l q_l \right) \end{aligned} \tag{11}$$

The chiral $SU_n * SU_n$ symmetry with the exact $SU_k * SU_k$ sub-symmetry is broken by the rotation of the $SU_k * SU_k$ invariant Hamiltonian density around the axis with the index $m = (n-1)^2 + 2k - 1$.

$$H_{SB} = e^{-2i\alpha Q^m}\ H_E\ e^{2i\alpha Q^m} \tag{12}$$

Only the quarks q_k and q_n are mixed. The $SU_k * SU_k$ symmetry is no longer exact. Only the term $\overline{q}_n q_n$ is rotated under transformation (12), because there is no $\overline{q}_k q_k$ term in the Hamiltonian density (11).

$$e^{-2i\alpha Q^m} H_E e^{2i\alpha Q^m} = \overline{q}_n q_n - \left(\overline{q}_n q_n - \overline{q}_k q_k \right)\ \sin^2\alpha - \frac{1}{2} \left(\overline{q}_k q_n + \overline{q}_n q_k \right)\ \sin 2\alpha \tag{13}$$

The above consideration is limited to processes not having the change of the quantum number N connected with the SU_n symmetry. So in the broken Hamiltonian density $H_{SB(\Delta N=0)}$ the terms $\overline{q}_n q_k$ and $\overline{q}_k q_n$ do not appear. The broken Hamiltonian density is a linear combination of the diagonal operators u^i only.

$$H_{SB(\Delta N=0)} = H_E + A \sum_{j=k}^{n-1} \left(\sqrt{\frac{j+1}{2j}} u^{(j+1)^2-1} - \sqrt{\frac{j-1}{2j}} u^{j^2-1} \right) \sin^2\alpha \tag{14}$$

where

$$A = \sqrt{2n}\ c_0 + (n-1) \sum_{j=k+1}^{n-1} c_{j^2-1} \sqrt{\frac{2}{j(j-1)}} \tag{15}$$

In more detail Eq. (14) is given as follows:

$$H_{SB(\Delta N=0)} = c_0\ u^0 - A\sqrt{\frac{k-1}{2k}}\sin^2\alpha\ u^{k^2-1} + \sum_{j=k+1}^{n-1} u^{j^2-1}\left(c_{j^2-1} + A\ \sin^2\alpha\ \sqrt{\frac{1}{2j(j-1)}}\right) +$$

$$-u^{n^2-1}\left(c_0\ \sqrt{n-1}\ + \sum_{j=k+1}^{n-1} c_{j^2-1}\ \sqrt{\frac{n(n-1)}{j(j-1)}}\right) - A\ u^{n^2-1}\ \sqrt{\frac{n}{2(n-1)}}\ \sin^2\ \alpha \tag{16}$$

If the $SU_k * SU_k$ symmetry is exact then the pseudo-scalar mesons corresponding to the indices $\left(j = 1, \ldots, k^2-1\right)$ are massless [9]. After the $SU_k * SU_k$ has been broken, the $SU_{k-1} * SU_{k-1}$ symmetry is still exact, because the operator Q^m does not mix the quarks $q_1, \ldots, q_{k-1}$ neither with themselves nor with other quarks. The mesons corresponding to the indices $\left(j = 1, \ldots, (k-1)^2 - 1\right)$ after symmetry breaking are still massless, while the mesons corresponding to the indices $\left(j = (k-1)^2, \ldots k^2 - 1\right)$ belong to the massive multiplet (k)[1] The masses of mesons are determined in the GMOR model. Before the $SU_n * SU_n$ symmetry is broken the masses are described by the coefficients $c_0, \ldots, c_{n^2-1}$ from Eq. (1). After the symmetry has been broken the new factors $c'_0, \ldots, c'_{n^2-1}$ are obtained as the coefficients standing by the operators u^i in the broken Hamiltonian density (16) [7].

$$c'_0 = c_0 \tag{17}$$

$$c'_{k^2-1} = A\sqrt{\frac{k-1}{2k}}\sin^2\alpha$$

$$c'_{j^2-1} = c_{j^2-1} + A\sin^2\alpha\sqrt{\frac{1}{2j(j-1)}} \quad (k<j<n)$$

$$c'_{n^2-1} = -c_0\ \sqrt{n-1} - \sum_{j=k+1}^{n-1} c_{j^2-1}\ \sqrt{\frac{n(n-1)}{j(j-1)}} + A\ \sqrt{\frac{n}{2(n-1)}}\ \sin^2\alpha$$

The masses of the mesons are determined as follows [12].

$$m_a^2 f_a^2 \delta_{ab} = \sqrt{\frac{2}{n}}\left(c_0\ d_{0ab} + \sum_{j=k+1}^{n} c'_{j^2-1} d_{(j^2-1)ab}\right) <u^0>_0 \tag{18}$$

The relation between the indices a, b, j and meson states is described, for example, for the $SU_4 * SU_4$ symmetry in [12, 13]. For $a = b = k^2 - l$, the mass of the (k) meson is given as follows

$$m_{(k)}^2 f_{(k)}^2 = \sqrt{\frac{2}{n}}\left(\sqrt{\frac{2}{n}}c_0\ + \sum_{j=k+1}^{n} d_{(j^2-1)ab} c'_{j^2-1}\right) <u^0>_0 = \frac{1}{k}\ \sqrt{\frac{2}{n}}\ A\sin^2\alpha\ <u^0>_0 \tag{19}$$

For $a = b = m = (n-1)^2 + 2k - 1$, the mass of the (n) meson is given by

$$m_{(n)}^2 f_{(n)}^2 = \sqrt{\frac{2}{n}}\left(\sqrt{\frac{2}{n}}c_0\ + \sum_{j=k+1}^{n} d_{(j^2-1)mm}\ c'_{j^2-1}\right) <u^0>_0 \tag{20}$$

Because

$$d_{(j^2-1)mm} = \sqrt{\frac{1}{2j(j-1)}} \quad (j<n) \tag{21}$$

$$d_{(n^2-1)mm} = \frac{2-n}{\sqrt{2n(n-1)}} \tag{22}$$

$$m^2_{(n)}f^2_{(n)} = \sqrt{\frac{1}{2n}}\, A\left(1-\left(1-\frac{1}{k}\right)\sin^2\alpha\right) < u^0 >_0 \tag{23}$$

In formulas (19) and (23) to determine the masses of (k) and (n) mesons one has (n-k + 3) unknown quantities with which to deal ($<u^0>_0, c_0,$ $c_{(k+1)^2-1}, \ldots, c_{n^2-1}, \sin\alpha$). Nevertheless the angle a is determined by the masses and decay constants of two pseudo-scalar mesons (k) and (n) only.

$$\sin^2\alpha = \frac{k\ m^2_{(k)}f^2_{(k)}}{2\ m^2_{(n)}f^2_{(n)} + (k-1)\ m^2_{(k)}f^2_{(k)}} \tag{24}$$

All the cases of symmetry breaking considered in [4–8] can be described by formula (24). Let us give simple examples: a) for k = 2, n = 3 a is the original Cabibbo angle Θ associated with rotation around the seventh axis in SU_3 subspace [4–7].

$$\sin^2\Theta = \frac{2\ m^2_{\pi}f^2_{\pi}}{2\ m^2_K f^2_K +\ m^2_{\pi}f^2_{\pi}} \tag{25}$$

b) for k = 2, n = 4 and rotation around the tenth axis [6–7] one obtains

$$\sin^2\alpha = \frac{2\ m^2_{\pi}f^2_{\pi}}{2\ m^2_D f^2_D +\ m^2_{\pi}f^2_{\pi}} \tag{26}$$

c) for k = 3, n = 4 and rotation around the fourteenth axis [7–8] one obtains

$$\sin^2\alpha = \frac{3\ m^2_K f^2_K}{2(m^2_D f^2_D +\ m^2_K f^2_K)} \tag{27}$$

In general the determination of the rotation angle (24) in $SU_n * SU_n$ symmetry is possible only if the new quantum numbers introduced by a transition to the higher symmetry are scalars of the charge type (additiv). So the Hamiltonian density (1) can be constructed as a linear combination of the diagonal operators u^i; only $\left(i = j^2-1, j = 1, \ldots, n\right)$. The method of determining the rotation angle, discussing and interpreting the symmetry breaking is described in more detail in [7] on $SU_4 * SU_4$ symmetry as an example.

2. Quarks mixing and the Cabibbo angle in the $SU_4 * SU_4$ broken symmetry

It is known that the Cabibbo angle has been introduced into SU_3 symmetry to explain the suppression of processes in which strangeness is not conserved [4]. The

Cabibbo angle is connected with the mixing of d and s quarks for weak interactions of hadrons. Its value, calculated by Oakes, does not contradict the experimental data. Before the charmed particles were discovered Glashow, Iliopoulos and Maiani [11] have suggested the generalization of a strong interaction symmetry to SU_4 [6]. The charged weak current is then given as follows

$$J_\mu = \overline{q}\ \gamma_\mu\ (1-\gamma_5)\ A\ q \tag{28}$$

where

$$A = \begin{pmatrix} 0 & 0 & \cos\Theta & \sin\Theta \\ 0 & 0 & -\sin\Theta & \cos\Theta \\ 0 & 0 & 0 & 0 \\ 0 & 0 & 0 & 0 \end{pmatrix} \tag{29}$$

The current (28) can be expressed in another form

$$J_\mu = (\overline{u}, \overline{c})\ \gamma_\mu\ (1-\gamma_5) \begin{pmatrix} \cos\Theta & \sin\Theta \\ -\sin\Theta & \cos\Theta \end{pmatrix} \begin{pmatrix} d \\ s \end{pmatrix} \tag{30}$$

so, quark mixing is described by an orthogonal matrix. On the grounds of Eq. (30) we cannot come to a conclusion about quarks in which the doublets are mixed. If the matrix A is generalized to the following form

$$A = \begin{pmatrix} 0 & 0 & \cos\Theta & \sin\Theta \\ 0 & 0 & -\sin\Theta & \cos\Theta \\ \cos\phi & \sin\phi & 0 & 0 \\ -\sin\phi & \cos\phi & 0 & 0 \end{pmatrix} \tag{31}$$

the quarks in the doublets (u, c) and (d, s) are mixed independently. The zeros in Eq. (31) are associated with the fact that the neutral currents which change the strangeness and/or charm are not observed. So, the current (28) can be given in the following form

$$J_\mu = (\overline{u}, \overline{c})\ \gamma_\mu\ (1-\gamma_5) \begin{pmatrix} \cos(\Theta+\phi) & \sin(\Theta+\phi) \\ -\sin(\Theta+\phi) & \cos(\Theta+\phi) \end{pmatrix} \begin{pmatrix} d \\ s \end{pmatrix} \tag{32}$$

If the currents only are taken into consideration we cannot solve the problem if the quarks are mixed in one or both doublets. This is not unexpected because the currents are built as a bi-linear combination of quark states and the angles Θ and ϕ, can always be substituted the effective angle $(\Theta+\phi)$. To solve the problem the Gell-Mann, Oakes, Renner (GMOR) model [9] will be used.

The charged weak current in SU_3 symmetry can be written as follows

$$J_\mu(\Theta) = \cos\Theta\ \left(J_\mu^1 + i\ J_\mu^2\right) + \sin\Theta\ \left(J_\mu^4 + i\ J_\mu^5\right) \qquad \Theta - \text{Cabibbo angle} \tag{33}$$

$$J_\mu = \overline{q}\ \gamma_\mu\ (1-\gamma_5)\lambda^k\ q \qquad q = \begin{pmatrix} u \\ d \\ s \\ c \end{pmatrix} \tag{34}$$

The current (33) can be obtained from the isospin component of the current $\left(J_\mu^1 + i\ J_\mu^2\right)$ by rotation through an angle 2Θ about the seventh axis in SU_3 space according to

$$J_\mu(\Theta) = e^{-2i\Theta F^7}\left(J_\mu^1 + i\ J_\mu^2\right)e^{2i\Theta F^7} \tag{35}$$

where

$$F^k = \int d^3x\ q^+(x)\ \frac{\lambda^k}{2}\ q(x) \tag{36}$$

The charged weak current in SU_4 symmetry (30) can be expressed in the following form

$$\begin{aligned} J_\mu(\Theta) = {} & \cos\Theta\ \left(J_\mu^1 + i\ J_\mu^2\right) + \sin\Theta\ \left(J_\mu^4 + i\ J_\mu^5\right) - \sin\Theta\ \left(J_\mu^{11} - i\ J_\mu^{12}\right) \\ & + \cos\Theta\ \left(J_\mu^{13} - i\ J_\mu^{14}\right) \end{aligned} \tag{37}$$

The current (37) can be obtained by rotation of the components $\Delta S = \Delta C$ through an angle 2Θ about the seventh axis in SU_4 space

$$J_\mu(\Theta) = e^{-2i\Theta F^7}\left(J_\mu^1 + i\ J_\mu^2 + J_\mu^{13} - i\ J_\mu^{14}\right)\ e^{2i\Theta F^7} \tag{38}$$

The transformation (38) changes the strangeness but not the charm because

$$\left[F^7, q_1\right] = \left[F^7, q_4\right] = 0 \tag{39}$$

The transformation (38) is connected with the mixing of d and s quarks (as in the case of SU_3 symmetry). In the SU_4 symmetry the mixing in electric charge subspace +2/3 can be taken into consideration. This is not possible in the SU_3 symmetry where only one state with the +2/3 charge exists. The possibility of expressing the current (37) by the transformation which changes charm but not strangeness should exist. The transformation has been described by Ebrahim in [6].

$$J_\mu(\phi) = e^{-2i\phi F^{10}}\left(J_\mu^1 + i\ J_\mu^2 + J_\mu^{13} - i\ J_\mu^{14}\right)\ e^{2i\phi F^{10}} \tag{40}$$

$$\left[F^{10}, q_2\right] = \left[F^{10}, q_3\right] = 0 \tag{41}$$

The transformation (40) is connected with the mixing of u and c quarks. The fact that there exist two transformations giving the current (37) but connected with different generators of the SU_4 group changing strangeness or charm respectively suggests that independent mixing in both doublets is possible. It is known that the Cabibbo angle is connected with strangeness non-conservation in weak interactions. The formula describing the value of the Cabibbo angle has been obtained by Oakes [4] in the procedure of symmetry breaking. Namely the $SU_3 * SU_3$ symmetry in the limit of the exact $SU_2 * SU_2$ symmetry is broken. The $SU_2 * SU_2$ subsymmetry is no longer exact. The symmetry is broken by the rotation of the $SU_2 * SU_2$ invariant Hamiltonian density through angle 2Θ about the seventh axis. Then the pion becomes massive. The symmetry breaking is connected with the mixing of d and s quarks. The rotation angle Θ, as a measure of symmetry violation, is a function of the mass and the decay constant of the pion and of the mass and the decay constant of the kaon as well (it is connected with the mixing of the strange quark and the strangeness non-conservation). If the breaking of the chiral $SU_2 * SU_2$ symmetry, the mass of the pion, the Cabibbo angle as well as a strangeness and charm non-conservation have a common origin then it seems that as a result of $SU_4 * SU_4$ symmetry breaking in the limit of the exact $SU_2 * SU_2$ sub-symmetry by the rotation of the $SU_2 * SU_2$ invariant Hamiltonian density through an angle 2ϕ about the tenth axis the angle ϕ connected with the mixing of u and c quarks as a measure of a symmetry violation should be a function of the mass and decay constant of the pion (breaking of the $SU_2 * SU_2$ symmetry) and a function of the mass and decay constant of a charmed meson (charm non-conservation). The cases of the separate and then simultaneously mixing of quarks in the sub-spaces of electric charge will be considered below.

If the electromagnetic mass splitting of u-d quarks is neglected the Hamiltonian density breaking the chiral $SU_4 * SU_4$ symmetry is given in the form

$$H = c_0 u^0 + c_8 u^8 + c_{15} u^{15} \tag{42}$$

where c_0, c_8, c_{15} are constants, u^a (a = 0, 1, ..., 15) are the scalar components of the $(\overline{4}, 4) + (4, \overline{4})$ representation of the chiral $SU_4 * SU_4$ group. On the grounds of the GMOR model the following relation for masses of the pseudo-scalar mesons can be obtained [12].

$$\begin{aligned} i \ <0|\left[\overline{Q}^a, \overline{D}^b\right]|0> &= \delta^{ab} f_a^2 m_a^2 + \int \frac{dq^2}{q^2} \rho^{ab} = \\ &= <u^0>_0 \left(\frac{c_0}{2} \delta^{ab} + \frac{c_8}{\sqrt{2}} d_{a8b} + \frac{c_{15}}{\sqrt{2}} d_{a15b} \right) + \\ &+ <u^8>_0 \left(\frac{c_0}{\sqrt{2}} d_{a8b} + c_8 d_{a8c} d_{b8c} + c_{15} d_{a8c} d_{b15c} \right) + \\ &+ <u_{15}>_0 \left(\frac{c_0}{\sqrt{2}} \ d_{a15b} + c_8 \ d_{a15c} \ d_{b8c} + c_{15} \ d_{a15c} \ d_{b15c} \right) \end{aligned} \tag{43}$$

where

$$\rho^{ab} = (2\pi)^3 \sum_{n \neq a} \delta^4 (p_n - q) < 0|\overline{D}^a|n> <n|\overline{D}^b|0> \tag{44}$$

f_a - decay constants, $<u^i>_0$ - vacuum expectation value of the operator u^i. Because the vacuum expectation values of operators u^8, u^{15} and the spectral density δ^{ab} are proportional to the squared parameters of symmetry breaking, they are further neglected [12]. Approximately from Eq. (43) we obtain

$$m_a^2 \ f_a^2 \ \delta^{ab} = \frac{1}{\sqrt{2}} \left(\frac{c_0}{\sqrt{2}} + c_8 \ d_{a8b} + c_{15} \ d_{a15b} \right) <u^0>_0 \tag{45}$$

The masses of the mesons are given as follows

$$\begin{aligned} m_\pi^2 \ f_\pi^2 &= \frac{1}{2\sqrt{3}} \left(\sqrt{3} \ c_0 + \sqrt{2} \ c_8 + c_{15} \right) <u^0>_0 \\ m_K^2 \ f_K^2 &= \frac{1}{2\sqrt{3}} \left(\sqrt{3} \ c_0 - \frac{1}{\sqrt{2}} \ c_8 + c_{15} \right) <u^0>_0 \end{aligned} \tag{46}$$

$$m_D^2\ f_D^2 = \frac{1}{2\sqrt{3}}\left(\sqrt{3}\ c_0 + \frac{1}{\sqrt{2}}\ c_8 - c_{15}\right) < u^0 >_0$$

In the limit of the exact chiral $SU_2 * SU_2$ sub-symmetry there is the following constraint

$$\sqrt{3}\ c_0 + \sqrt{2}\ c_8 + c_{15} = 0 \tag{47}$$

so the pion is massless.

Let us make some remarks. The task of the Cabibbo angle calculation in SU_4 symmetry using the procedure of symmetry breaking has been done in [6]. In Ebrahim's earlier paper [5] the parameters of the $SU_4 * SU_4$ symmetry breaking have been found.

$$\frac{c_8}{c_0} = -\frac{2\sqrt{2}}{\sqrt{3}}\frac{m_K^2 f_K^2 - m_\pi^2 f_\pi^2}{m_K^2 f_K^2 + m_D^2 f_D^2} \qquad \frac{c_{15}}{c_0} = -\frac{1}{\sqrt{3}}\frac{3m_D^2 f_D^2 - m_K^2 f_K^2 - 2m_\pi^2 f_\pi^2}{m_K^2 f_K^2 + m_D^2 f_D^2} \tag{48}$$

In [6] the numerical values of parameters (48) have been used to calculate the rotation angle (interpreted as the Cabibbo angle). The $SU_2 * SU_2$ invariant Hamiltonian density breaking $SU_4 * SU_4$ symmetry has been rotated through an angle 2 Θ about the seventh axis and the coefficients of the operators u^a (a = 0, 8, 15) have been identified with the parameters of symmetry breaking

$$\begin{aligned} H_{SB}(\Delta S = 0) = c_0 u^0 + \frac{\sqrt{3}}{2} c_8 \sin^2\Theta\ u^3 + c_8\left(1 - \frac{3}{2}\sin^2\Theta\right)\ u^8 \\ - \left(\sqrt{3}c_0 + \sqrt{2}c_8\right)\ u^{15} \end{aligned} \tag{49}$$

It seems to us that there are some errors in the numerical calculations of the author. The use of the numerical values of the parameters (48) has not been necessary. On the grounds of theoretical formulas only, indeed from the Eq. (7) in Ref. [5] and the Eq. (10) in A3-Ebrahim, it follows that

$$\sin^2\Theta = \frac{2m_\pi^2 f_\pi^2}{2m_K^2 f_K^2 + m_\pi^2 f_\pi^2} \tag{50}$$

Then the value of Θ is given by

$$\sin^2\Theta = (0.215)^2 \tag{51}$$

instead of

$$\sin^2\Theta = -0.04 \tag{52}$$

from Eqs. (10) in [6]. Formula (50) has the same form as in SU_3 symmetry. In agreement with our expectation the angle Θ is described by parameters of the pion and the strange meson.

In Ebrahim's method the $SU_4 * SU_4$ symmetry breaking the Hamiltonian density is parametrized by the factors c_0, c_8, c_{15}. The parameters of symmetry breaking are expressed by the masses and decay constants of the mesons and they are fixed (Eq. (7) in [5]). In the limit of the exact $SU_2 * SU_2$ sub-symmetry the factors c_0, c_8, c_{15} should satisfy the constraint (47) but it is possible only if $m_\pi = 0$ namely the parameters of symmetry breaking are not expressed by real (measured in experiment) masses of mesons. In [6] Ebrahim breaks the $SU_4 * SU_4$ symmetry in the limit of the exact $SU_2 * SU_2$ sub-symmetry by the rotation of the $SU_2 * SU_2$ invariant Hamiltonian density through an angle 2 Θ about the seventh axis. The factors of the rotated Hamiltonian density are identified with the parameters of symmetry breaking (Eq. (7) in [5]). Solving a set of equations the author gets the factors c_0, c_8, c_{15} dependent on the rotation angle and on the real mesons masses already. The masses of mesons standing in the formula which describes the parameters of symmetry breaking are determined by the method of symmetry breaking and they have a real value for the real realization of the symmetry breaking only. In this case the rotation angle does not matter a parameter of the symmetry violation. It seems to us that such an interpretation is not satisfactory. The expression of meson masses as a function of the rotation angle (as a measure of symmetry violation) seems to be more natural. In the present paper the other interpretation of the symmetry breaking and the method of calculating the rotation angle is proposed. We describe our method as follows.

Before the $SU_4 * SU_4$ symmetry in the limit of the exact $SU_2 * SU_2$ sub-symmetry is broken the masses of mesons have been expressed by the factors c_0, c_8, c_{15} which satisfy the constraint (47). After symmetry breaking a new set of factors c'_0, c'_g, c'_{15} dependent on the old factors c_0, c_8, c_{15} and on the rotation angle is introduced. The new factors are identified with the coefficients by the operators u^i of the rotated Hamiltonian density (49).

$$c'_0 = c_0 \qquad c'_g = c_8\left(1 - \frac{3}{2} sin^2\Theta\right) \qquad c'_{15} = -\sqrt{3}c_0 - \sqrt{2}c_8 \tag{53}$$

Meson masses are expressed by new factors and they are the function of the rotation angle as a measure of symmetry violation.

$$m_\pi^2 f_\pi^2 = \frac{1}{2\sqrt{3}}\left(\sqrt{3}c'_0 + \sqrt{2}c'_g + c'_{15}\right) < u^0 >_0 = -\frac{\sqrt{3}}{2\sqrt{2}} c_8 \sin^2\Theta < u^0 >_0 \tag{54}$$

$$m_K^2 f_K^2 = \frac{1}{2\sqrt{3}}\left(\sqrt{3}c'_0 - \frac{c'_8}{\sqrt{2}} + c'_{15}\right) < u^0 >_0 = -\frac{\sqrt{3}}{2\sqrt{2}} c_8\left(1 - \frac{1}{2}\sin^2\Theta\right) < u^0 >_0$$

$$m_D^2 f_D^2 = \frac{1}{2\sqrt{3}}\left(\sqrt{3}c'_0 + \frac{c'_8}{\sqrt{2}} - c'_{15}\right) < u^0 >_0 = \left(c_0 + \frac{\sqrt{3}}{2\sqrt{2}} c_8\left(1 - \frac{1}{2}\sin^2\Theta\right)\right) < u^0 >_0$$

It seems to be more natural that the meson masses are functions of the parameters of symmetry breaking (54) than inversely the parameters of symmetry breaking are functions of meson masses which are not consistent with the experimental data and are dependent on the method of symmetry breaking. This interpretation is consistent with the fact that the mass generation of the mesons is a consequence of symmetry breaking. From Eq. (54) we obtain the formula for the angle Θ as in Eq. (50). Let us consider the other variant of symmetry breaking described in [6]. Ebrahim, using his method, broke the $SU_4 * SU_4$ symmetry in the limit of the exact $SU_3 * SU_3$ symmetry by the rotation of $SU_3 * SU_3$ invariant Hamiltonian density about the fourteenth axis in the SU_4 space. The rotation angle Θ' is identified with the Cabibbo angle. The formula describing the angle Θ' should be given as follows

$$\sin^2\Theta' = \frac{3m_K^2 f_K^2}{2(m_K^2 f_K^2 + m_D^2 f_D^2)} \tag{55}$$

(in Eqs. (4a) in [8] there is the factor 3/2). The rotation of the Hamiltonian density about the fourteenth axis is considered in [14] too. The D meson is interpreted as a Goldstone boson. Putting aside the agreement of the numerical value of the angle Θ' with the experimental data it seems to us that the angle connected with the rotation about the fourteenth axis cannot be interpreted as the Cabibbo angle, because the rotation is performed inside the doublet (s, c). Then the states with the different electric charges are mixed. The interpretation that the D meson is a Goldstone boson is also unsatisfactory. If the $SU_4 * SU_4$ symmetry is broken in such a way that the $SU_2 * SU_2$ sub-symmetry is still exact, so the K meson becomes massive but the pion is still massless. Such a symmetry breaking cannot be accepted, results contradict the experimental data. The next breaking of the exact $SU_3 * SU_3$ symmetry is connected with the mixing of s and c quarks. The rotation angle cannot be interpreted as the Cabibbo angle for the reasons given above. It seems that the hierarchy of symmetry breaking is extended and the breaking of the $SU_4 * SU_4$ symmetry taken as a whole cannot be connected with the Cabibbo angle. This is possible, however, for $SU_4 * SU_4$ symmetry breaking in the limit of exact $SU_2 * SU_2$ sub-symmetry. Then results are in agreement with our expectation.

Our method described above is used· to calculate an angle ϕ which is connected with the rotation about the tenth axis in SU_4 space. Then the $SU_4 * SU_4$ symmetry is broken by the rotation of the $SU_2 * SU_2$ invariant Hamiltonian density through an angle 2ϕ about the tenth axis.

$$H_{SB}(\Delta C = 0) = c_0 u^0 + \left(\sqrt{2}c_0 + \frac{\sqrt{3}}{2}c_8\right)\sin^2\phi\ u^3 + \left(c_8\ \left(\frac{1}{2\sqrt{2}}\left(\frac{4}{\sqrt{3}}c_0 + \sqrt{2}c_8\right)\sin^2\phi\right)\ u^8 + \right.$$
$$+\left(-\sqrt{3}\ c_0 + \sqrt{2}\ c_8 + \left(\frac{4}{\sqrt{3}}c_0 + \sqrt{2}c_8\right)\sin^2\phi\right)\ u^{15} \tag{56}$$

Using the factors from the Hamiltonian density (56) the masses of mesons are given as follows

$$m_\pi^2 f_\pi^2 = \left(c_0 + \frac{\sqrt{6}}{4}\right)\sin^2\phi\ \ <u^0>_0 \tag{57}$$

$$m_K^2 f_K^2 = -\left(\frac{\sqrt{6}}{4} - \frac{1}{2}\left(c_0 + \frac{\sqrt{6}}{4}c_8\right)\sin^2\phi\right) <u^0>_0$$

$$m_D^2 f_D^2 = \left(1 - \frac{1}{2}\sin^2\phi\right)\left(c_0' + \frac{\sqrt{6}}{4}c_8\right) <u^0>_0$$

so

$$\sin^2\phi = \frac{2m_\pi^2 f_\pi^2}{2m_D^2 f_D^2 + m_\pi^2 f_\pi^2} \tag{58}$$

In agreement with our expectation the angle ϕ is a function of the mass of the pion (as a measure of the $SU_2 * SU_2$ violation) and is connected with the parameters of the charmed meson (mixing in (u, c) doublet). For the mass m_D = 1862 MeV and f_D/f_π = 0.974 [13] one gets

$$\sin^2\phi = 0.076 \tag{59}$$

The small value of the angle ϕ is the effect of the large mass of the charmed quark. From (59) results that only mixing in the (u, c) system is excluded, the value of the angle ϕ contradicts the experimental data. The simultaneous mixing in both doublets are, however, still possible. Fritzsch [15] considers also the mixing in (u, c) system. The mixing angle is calculated on the grounds of quark masses and does not contradict the results obtained above. Although the value of the angle ϕ is relatively small, it is significant: the sum of the angles $\Theta + \phi$ is larger than the value of the angle measured experimentally, called Cabibbo angle. This fact cannot be explained by the limits of experimental errors. Let us note that the angles (50) and (58) are calculated for the case where quarks are mixed separately. The angles from formula (32) cannot be identified with those from Eqs. (50) and (58). In the case of simultaneous mixing in both doublets the relation between the angles is more complicated. To find the relation, the $SU_2 * SU_2$ invariant Hamiltonian density is rotated through an angle 2ϕ about the tenth axis and afterwards by an angle 2Θ about the seventh axis. The sequence of the rotations is insignificant, because

$$\left[F^7, F^{10}\right] = 0 \tag{60}$$

The rotated Hamiltonian density is given by

$$H_{SB}(\Delta S = \Delta C = 0) = c_0 u^0 + \left(\frac{\sqrt{3}}{2}\ c_8\ \sin^2\Theta + \left(\sqrt{2}\ c_0 + \frac{\sqrt{3}}{2} c_8\right) \sin^2\phi\right) u^3 + \tag{61}$$

$$+\left(\frac{1}{\sqrt{3}}\left(\sqrt{2} c_0 + \frac{\sqrt{3}}{2} c_8\right) \sin^2\phi + c_8\left(1 - \frac{3}{2} \sin^2\Theta\right)\right)\ u^8 +$$

$$+\left(-\sqrt{3}\ c_0 - \sqrt{2}\ c_8 + \frac{2\sqrt{2}}{\sqrt{3}}\left(\sqrt{2} c_0 + \frac{\sqrt{3}}{2} c_8\right) \sin^2\phi\right)\ u^{15}$$

The meson masses are given as follows

$$m_\pi^2 f_\pi^2 = \frac{1}{2\sqrt{3}}\left(\sqrt{6}\left(\sqrt{2}\ c_0 + \frac{\sqrt{3}}{2}\ c_8\right) \sin^2\phi - \frac{3}{\sqrt{2}}\ c_8\ \sin^2\Theta\right)\ <u^0>_0 \tag{62}$$

$$m_K^2 f_K^2 = \frac{1}{2\sqrt{3}}\left(\frac{3}{\sqrt{6}}\left(\sqrt{2}\ c_0 + \frac{\sqrt{3}}{2}\ c_8\right) \sin^2\phi - \frac{3}{\sqrt{2}}\ c_8\ \left(1 - \frac{1}{2} \sin^2\Theta\right)\ <u^0>_0\right.$$

$$m_D^2 f_D^2 = \frac{1}{2\sqrt{3}}\left(2\sqrt{3}\ c_0 - \frac{3}{\sqrt{6}}\ \left(\sqrt{2} c_0 + \frac{\sqrt{3}}{2} c_8\right) \sin^2\phi + \frac{3}{\sqrt{2}}\ c_8\ \left(1 - \frac{1}{2} \sin^2\Theta\right)\right)\ <u^0>_0$$

Now the angles Θ and ϕ cannot be described independently. The following relation is obeyed.

$$2\ m_\pi^2 f_\pi^2 + 2\ \left(m_K^2 f_K^2 + m_D^2 f_D^2\right)\ \sin^2\Theta\ \sin^2\phi = \tag{63}$$

$$= \left(2\ m_K^2 f_K^2 + m_\pi^2 f_\pi^2\right)\ \sin^2\Theta + \left(2\ m_D^2 f_D^2 + m_\pi^2 f_\pi^2\right)\ \sin^2\phi$$

or equivalently

$$1 + \left(\frac{1}{\sin^2\Theta} + \frac{1}{\sin^2\phi} - 1\right) \sin^2\Theta\ \sin^2\phi = \frac{\sin^2\Theta}{\sin^2\Theta_0} + \frac{\sin^2\phi}{\sin^2\phi_0} \tag{64}$$

where

$$\sin^2\Theta_0 = \frac{2m_\pi^2 f_\pi^2}{2m_K^2 f_K^2 + m_\pi^2 f_\pi^2} \qquad \sin^2\phi_0 = \frac{2m_\pi^2 f_\pi^2}{2m_D^2 f_D^2 + m_\pi^2 f_\pi^2} \tag{65}$$

The angles Θ and ϕ from Eq. (64) concern a simultaneous mixing in doublets (d, s) and (u, c) respectively and they can be identified with those from Eq. (32). The condition (64) limits the values of the angles Θ and ϕ. The maximal values of the angles Θ_0 and ϕ_0 are given by Eq. (65). The value of the function

$$f(\Theta, \phi) = \sin(\Theta + \phi) \tag{66}$$

is also limited. A numerical calculation shows that there is an extremum (a maximum) of function (66) on the condition (64) for

$$\Theta_m = 0.20452 \qquad \phi_m = 0.02575 \qquad \sin(\Theta_m + \phi_m) = 0.2282 \tag{67}$$

It is worth noticing that the extremum of function (66) on condition (64) can be identified with the measured Cabibbo angle. It is not excluded that symmetry breaking is realized in the maximal allowed case, so the effective angle of mixing would correspond to the maximum of the function (66).

3. Bonds for the Kobayashi-Maskawa mixing parameters in a model with hierarchical symmetry breaking

A simultaneous mixing in (d, s) and (u, c) sectors has also been taken into account [15, 16], but due to the large mass of the c quark, the influence of the mixing in the (u, c) sector can be treated as a perturbation. At the six-quark level the quark mixing is described by three Cabibbo-like flavor mixing angles and the phase parameter responsible for CP-non-conservation [17]. The charged weak current in the $SU_6 * SU_6$ chiral symmetry

$$J_\mu = (\overline{u}, \overline{c}, \overline{t})\ \gamma_\mu\ (1 - \gamma_5)\ U \begin{pmatrix} d \\ s \\ b \end{pmatrix} \tag{68}$$

is described by a unitary matrix U, which can be put in 21 different forms [18], however only the standard Kobayashi-Maskawa matrix [19] will be used further.

$$U = \begin{pmatrix} c_1 & s_1 c_3 & s_1 s_3 \\ -s_1 c_2 & c_1 c_2 c_3 - s_2 s_3 e^{i\delta} & c_1 c_2 s_3 + s_2 c_3 e^{i\delta} \\ s_1 s_2 & -c_1 s_2 c_3 - c_2 s_3 e^{i\delta} & -c_1 s_2 s_3 + c_2 s_3 e^{i\delta} \end{pmatrix} \tag{69}$$

where $s_i = \sin\Theta_i$, $c_i = \cos\Theta_i$.
The matrix (69) can be expressed as follows

$$U = \begin{pmatrix} 1 & 0 & 0 \\ 0 & c_2 & s_2 \\ 0 & -s_2 & c_2 \end{pmatrix} \begin{pmatrix} 1 & 0 & 0 \\ 0 & 1 & 0 \\ 0 & 0 & e^{i\delta} \end{pmatrix} \begin{pmatrix} c_1 & s_1 & 0 \\ -s_1 & c_1 & 0 \\ 0 & 0 & 1 \end{pmatrix} \begin{pmatrix} 1 & 0 & 0 \\ 0 & c_3 & s_3 \\ 0 & -s_3 & c_3 \end{pmatrix} \tag{70}$$

$$U = U_2\ U_\delta\ U_1\ U_3 \tag{71}$$

and it can mix quarks either in the negative or in the positive electric charge subspace.

A simultaneous mixing in both spaces was also considered [10]. From the form of the matrix (70) the following variants of the quark mixing are allowed:

$$A: \quad U = U_2(s-b)\ U_\delta\ U_1(d-s)\ U_3(s-b) \tag{72}$$

$$B: \quad U = U_2(c-t)\ U_\delta\ U_1(d-s)\ U_3(s-b) \tag{73}$$

$$C: \quad U = U_2(c-t)\ U_\delta\ U_1(u-c)\ U_3(s-b) \tag{74}$$

$$D: \quad U = U_2(c-t)\ U_\delta\ U_1(u-c)\ U_3(c-t) \tag{75}$$

where $U_k(x-y)$ denotes the mixing of x and y quarks by the matrix U_k. It is known that the Cabibbo angle cannot be explained by the mixing in the (u-c) sector only [15, 16], so the variants C and D must be rejected. Let us examine the variant B.

The charged weak current (68) with the matrix (69) for the variant B can be expressed as follows

$$J_\mu = R\ J_\mu(0)\ R^{-1} \tag{76}$$

where

$$J_\mu(0) = (\overline{u}, \overline{c}, \overline{t})\ \gamma_\mu\ (1-\gamma_5)\ I \begin{pmatrix} d \\ s \\ b \end{pmatrix} \tag{77}$$

$$R = e^{-2i\Theta_3 Q^{21}} e^{-2i\Theta_1 Q^7} e^{-i\delta X} e^{2i\Theta_2 Q^{32}} \tag{78}$$

$$X = \frac{4}{\sqrt{10}} Q^{24} - \frac{1}{\sqrt{15}} Q^{35} \tag{79}$$

where Q^k is the $6*6$ matrix representation of the k-th generator of SU_6 group. To get the values of the angles Θ_i the Gell-Mann-Oakes-Renner model will be used [9]. If the electromagnetic mass splitting of u-d quarks is neglected the Hamiltonian density breaking the chiral $SU_6 * SU_6$ symmetry is given as follows

$$H_0 = c_0 u^0 + c_8 u^8 + c_{15} u^{15} + c_{24} u^{24} + c_{35} u^{35} \tag{80}$$

where $c_0, \ldots, c_{35}$ are the symmetry breaking parameters, u^i (i = 0,1, ..., 35) are the scalar components Of the $(\overline{6}, 6) + (6, \overline{6})$ representation of the chiral $SU_6 * SU_6$ group. From the GMOR model, neglecting the vacuum expectation values of operators u^k (k = 8, 15, 24, 35) and the spectral density ρ^{ab} as proportional to the squared parameters of the symmetry breaking [12, 16], we get the approximate relation for masses of the pseudo-scalar mesons

$$m_a^2\ f_a^2\ \delta^{ab} = \frac{1}{\sqrt{3}} \left(\frac{c_0}{\sqrt{3}} + c_8\ d_{a8b} + c_{15}\ d_{a15b} + c_{24}\ d_{a24b} + c_{35}\ d_{a35b} \right) < u^0 >_0 \tag{81}$$

where f_a are the decay constants, d_{aib} - symmetric constants of the SU_6 group, $< u^0 >_0$ - the vacuum expectation value of the operator u^0. From (81) we obtain

$$\pi = m_\pi^2\ f_\pi^2 \ = \frac{1}{\sqrt{3}}\left(\frac{c_0}{\sqrt{3}} + \frac{c_8}{\sqrt{3}} + \frac{c_{15}}{\sqrt{6}} + \frac{c_{24}}{\sqrt{10}} + \frac{c_{35}}{\sqrt{15}}\right) \ < u^0 >_0 \tag{82}$$

$$K = m_K^2\ f_K^2 \ = \frac{1}{\sqrt{3}}\left(\frac{c_0}{\sqrt{3}} - \frac{c_8}{2\sqrt{3}} + \frac{c_{15}}{\sqrt{6}} + \frac{c_{24}}{\sqrt{10}} + \frac{c_{35}}{\sqrt{15}}\right) \ < u^0 >_0 \tag{83}$$

$$D = m_D^2\ f_D^2 \ = \frac{1}{\sqrt{3}}\left(\frac{c_0}{\sqrt{3}} + \frac{c_8}{2\sqrt{3}} - \frac{c_{15}}{\sqrt{6}} + \frac{c_{24}}{\sqrt{10}} + \frac{c_{35}}{\sqrt{15}}\right) \ < u^0 >_0 \tag{84}$$

$$B = m_B^2\ f_B^2 \ = \frac{1}{\sqrt{3}}\left(\frac{c_0}{\sqrt{3}} + \frac{c_8}{2\sqrt{3}} + \frac{c_{15}}{2\sqrt{6}} - \frac{3c_{24}}{2\sqrt{10}} + \frac{c_{35}}{\sqrt{15}}\right) \ < u^0 >_0 \tag{85}$$

$$T = m_T^2\ f_T^2 \ = \frac{1}{\sqrt{3}}\left(\frac{c_0}{\sqrt{3}} + \frac{c_8}{2\sqrt{3}} + \frac{c_{15}}{2\sqrt{6}} + \frac{c_{24}}{2\sqrt{10}} - \frac{2c_{35}}{\sqrt{15}}\right) \ < u^0 >_0 \tag{86}$$

By the symmetry breaking, the massless quark x can become massive if it is mixed with the other massive y. The rotation angle is then described by the masses of pseudo-scalar mesons. If the $SU_n * SU_n$ symmetry with the exact $SU_k * SU_k$ sub-symmetry is broken to the exact $SU_{k-1} * SU_{k-1}$ symmetry, the rotation angle is a function of masses of a pseudo-scalar meson belonging to n-multiplet of the $SU_n * SU_n$ group and the meson which has become massive [19]. We demand the quarks to become massive due to the hierarchical symmetry breaking, so the highest exact symmetry of the Hamiltonian density, which Can be assumed, is $SU_4 * SU_4$ (at least one quark in the each sector must be massive). Oakes and the others [4, 20, 21] in order to get the Cabibbo angle value in the $SU_3 * SU_3$ or $SU_4 * SU_4$ symmetry, have rotated the Hamiltonian density breaking the chiral symmetry in the same way as the weak charged current. In a model with hierarchical symmetry breaking such a procedure cannot be used. Let us notice that from the form (5) of the rotation operator R it follows that the quarks are mixed in the following sequence: (c-t), a phase rotation, (d-s), (s-b), so for the exact $SU_4 * SU_4$ symmetry the massless quarks d and s would be mixed as the first (in the negative electric charge subspace) and then the generation of their masses would not be possible. The quark s would become massive in the next stage of the symmetry breaking after the mixing with the massive quark b. So, in order to get the massive both d and s quarks, they should be mixed in the inverse sequence. In the first stage of the symmetry breaking the exact $SU_4 * SU_4$ symmetry is broken to the exact $SU_2 * SU_2$ symmetry, in the 2nd stage even the $SU_2 * SU_2$ symmetry is no longer exact. The next mixing stages are connected either with the mass generation of the c quark (variant B) or with the repeated mixing of massive s and b quarks (variant A). In our procedure the Hamiltonian density breaking the chiral $SU_6 * SU_6$ symmetry will be rotated in the inverse sequence in comparison with the rotation of the weak charged current.

$$H_{SB} = R_1\ H_0\ R_1^{-1} \tag{87}$$

where

$$R_1 = e^{2i\ \Theta_2\ Q^{32}} e^{-iX\ \delta} e^{-2i\ \Theta_1\ Q^7} e^{-2i\ \Theta_3\ Q^{21}} \tag{88}$$

The exact $SU_4 * SU_4$ symmetry implies the following relations

$$c_8 = c_{15} = 0 \qquad \sqrt{5}\ c_0 + c_{35} = 0 \tag{89}$$

So, the $SU_4 * SU_4$ invariant Hamiltonian density is given as

$$H_E = c_0 \left(u^0 - \sqrt{5}\ u^{35}\right) + c_{24} \left(u^{24} - \sqrt{\frac{3}{2}}\ u^{35}\right) \tag{90}$$

or equivalently

$$H_E = P\ \overline{q}_6\ q_6 - V\ \overline{q}_5\ q_5 \tag{91}$$

where

$$P = \sqrt{c_0} + V \qquad V = \frac{5}{\sqrt{10}}\ c_{24} \tag{92}$$

The symmetry-breaking Hamiltonian density

$$H_{SB} = R_1\ H_E\ R_1^{-1} \tag{93}$$

retaining the flavor-conservation part only is given as follows

$$H_{SB} = \overline{q}_6 q_6 P c_2^2 - \overline{q}_5 q_5 V c_3^2 + \overline{q}_4 q_4 P s_2^2 - \overline{q}_3 q_3 V c_1^2 s_3^2 - \overline{q}_2 q_2 V s_1^2 s_3^2 \tag{94}$$

Let us notice that the phase transformation does not produce terms $\overline{q}_i q_i$ since the operator (79) commutes with the scalar components u^k. The flavor-conservation on each stage of the symmetry breaking has been assumed. The Hamiltonian density (94) can be written as a function of the operators if, so the coefficients of u^k's are given as

$$c_0' = c_0 \tag{95}$$

$$c_{34}' = \frac{V}{2}\ s_1^2\ s_3^2 \tag{96}$$

$$c_8' = \frac{V}{2\sqrt{3}}\ \left(2\ c_1^2\ s_3^2 - s_1^2\ s_3^2\right) \tag{97}$$

$$c_{15}' = -\frac{1}{2\sqrt{6}}\ \left(3\ P\ s_2^2 + V\ s_3^2\right) \tag{98}$$

$$c_{24}' = \frac{1}{2\sqrt{10}}\ \left(4\ V - 5\ V\ s_3^2 + P\ s_2^2\right) \tag{99}$$

$$c_{35}' = -\frac{1}{2\sqrt{15}}\ \left(5\ P + V - 6\ P\ s_2^2\right) \tag{100}$$

Now, after the symmetry breaking, the pseudo-scalar masses (82–86) will be described as functions of the coefficients c_i' (i = 0, 3, 8, 15, 24, 35) [7, 16].

$$\pi = Z\ V\ s_1^2\ s_3^2 \tag{101}$$

$$K = Z\ V\ s_3^2\ \left(1 - \frac{1}{2}\ s_1^2\right) \tag{102}$$

$$D = -Z\ \left(P\ s_2^2 - \frac{1}{2}\ V\ s_1^2\ s_3^2\right) \tag{103}$$

$$B = Z\ V\ \left(1 - s_3^2\ \left(1 - \frac{1}{2}\ s_1^2\right)\right) \tag{104}$$

$$T = -Z\left(P\ c_2^2 - \frac{1}{2}\ V\ s_1^2\ s_3^2\right) \tag{105}$$

where

$$Z = -\frac{1}{2\sqrt{3}}\ \ <u^0>_0 \tag{106}$$

The Cabibbo angle Θ_1 is expressed in the same form as at four-quark level in the $SU_4 * SU_4$ symmetry [7, 16],

$$s_1^2 = \frac{\pi}{2\ K + \pi} \tag{107}$$

Because

$$s_1^2\ s_3^2 = \frac{\pi}{K + B} \quad \Rightarrow \quad s_3^2 = \frac{K + \frac{\pi}{2}}{K + B} \tag{108}$$

In an agreement with our prediction the angle Θ_3 connected with the mixing of s and b quarks is expressed by the parameters of the strange and beautiful mesons. The angle Θ_1 however, connected with mixing d and s quarks and breaking of the $SU_2 * SU_2$ symmetry is expressed by the masses of the pion and the kaon. The angle Θ_2 connected with the mixing in the (c-t) sector is given as

$$s_2^2 = \frac{D - \frac{\pi}{2}}{D + T - \pi} \tag{109}$$

Let us notice that if we do not demand the flavor-conservation on each stage of the symmetry breaking, after the rotation around the 21st axis the terms $\overline{q}_5 q_3, \overline{q}_3 q_5$ in the broken Hamiltonian density arise. In the second stage (the rotation around the 7th axis) there will be in H_{SB} the following terms: $\overline{q}_5 q_3, \overline{q}_3 q_5, \overline{q}_2 q_3, \overline{q}_3 q_2, \overline{q}_5 q_2, \overline{q}_2 q_5$. Because

$$\left[X, \overline{q}_3\ q_5\right] = i\ \delta\ \overline{q}_3\ q_5 \qquad \left[X, \overline{q}_5\ q_3\right] = -i\ \delta\ \overline{q}_5\ q_3 \tag{110}$$

after the phase rotation there will arise in the H_{SB} the following terms: $\overline{q}_3\ q_5\ e^{i\delta}$, $\overline{q}_5\ q_3\ e^{-i\delta}$, ... In the variant B the matrix U_2 has mixed c and t quarks so that in the flavor-conservation part of the broken Hamiltonian density the phase factor $e^{i\delta}$ cannot appear. But if the matrix U_2 mixes s and b quarks again, due to the following relations

$$e^{-2i\Theta_2 Q^{21}}\overline{q}_3 q_5 e^{2i\Theta_2 Q^{21}} = \overline{q}_3 q_5 c_2^2 - \overline{q}_5 q_3 s_2^2 + \frac{1}{2}\left(\overline{q}_5 q_5 - \overline{q}_3 q_3\right)\ \sin 2\Theta_2 \tag{111}$$

$$e^{-2i\Theta_2,\ Q^{21}}\overline{q}_5 q_3 e^{2i\Theta_2 Q^{21}} = \overline{q}_5 q_3 c_2^2 - \overline{q}_3 q_5 s_2^2 + \frac{1}{2}\left(\overline{q}_5 q_5 - \overline{q}_3 q_3\right) \sin 2\Theta_2 \tag{112}$$

the following terms: $\overline{q}_5\ q_5\ e^{i\delta}, \overline{q}_5\ q_5\ e^{-i\delta}, \overline{q}_3\ q_3\ e^{i\delta}, \overline{q}_3\ q_3\ e^{-i\delta}$ appear in the broken Hamiltonian density.

We assume that the symmetry is broken by the quarks mixing in the following sequence: (s-b), (d-s), a phase rotation, (s-b) and the flavor will not be conserved in the intermediate stages of the symmetry breaking, but it will be conserved in the broken symmetry taken as a whole. The assumptions given above are consistent with the variant A. Let us take it into account.

We assume the exact $SU_4 * SU_4$ symmetry. The Hamiltonian density is given by Eq. (91). After symmetry breaking, the flavor conserving part of the broken Hamiltonian density H_{SB} is given as

$$H_{(\Delta F=0)} = \bar{q}_6 q_6 P - \bar{q}_5 q_5 V(\alpha - A) - \bar{q}_3 q_3 V(\beta + A) - \bar{q}_2 q_2 V\gamma \tag{113}$$

where

$$\alpha = c_1^2 s_2^2 s_3^2 + c_2^2 c_3^2 \tag{114}$$

$$\beta = c_1^2 c_2^2 s_3^2 + s_2^2 c_3^2 \tag{115}$$

$$\gamma = s_1^2 s_3^2 \tag{116}$$

$$A = \frac{1}{2} \cos\Theta_1 \sin 2\Theta_2 \sin 2\Theta_3 \cos\delta \tag{117}$$

Since

$$\alpha + \beta + \gamma = 1 \tag{118}$$

α can be eliminated from (113).

The coefficients by the operators u^k are as follows

$$c_0' = c_0 \tag{119}$$

$$c_3' = \frac{V}{2}\gamma \tag{120}$$

$$c_8' = \frac{V}{2\sqrt{3}}(2\beta + 2A - \gamma) \tag{121}$$

$$c_{15}' = -\frac{V}{2\sqrt{6}}(\beta + A + \gamma) \tag{122}$$

$$c_{24}' = \frac{V}{2\sqrt{10}}(4 - 5(\beta + A + \gamma)) \tag{123}$$

$$c_{35}' = -\sqrt{5}c_0 - \sqrt{\frac{3}{2}}c_{24} \tag{124}$$

Let us notice that the functions β and A occur in Eqs. (121–123) as a sum $\beta + A$ only. So, only two functions can be expressed independently. Because there is no mixing in the positive electric charge subspace we shall not use relations describing mesons D and T. The following relations are obeyed

$$\pi = ZV\gamma \tag{125}$$

$$K = ZV\left(\beta + A + \frac{\gamma}{2}\right) \tag{126}$$

$$B = ZV\left(1 - \beta - A - \frac{\gamma}{2}\right) \tag{127}$$

so we immediately obtain

$$s_1^2 s_3^2 = \frac{\pi}{K + B} \tag{128}$$

as in the variant B, but at the moment the angles Θ_1 and Θ_3 cannot be calculated separately. Putting the experimental value

$$\cos\Theta_1 = 0.9737 \qquad (\sin\Theta_1 = 0.2278) \tag{129}$$

as an input [22], we get

$$\sin\Theta_3 = 0.136 \qquad (\Theta_3 = 7.8^\circ) \tag{130}$$

for

$$m_\pi = 0.139 \ \ GeV \tag{131}$$

$$m_K = 0.495 \ \ GeV \qquad f_K = 1.28 \quad [21] \tag{132}$$

$$m_B = 5.2 \ \ GeV \qquad f_B = 0.86 \quad [22] \tag{133}$$

The angle Θ_3 was calculated by Fritzsch [15] also for the following quark masses ratios:

$$m_u \ : \ m_d \ : \ m_s \ : \ m_c = 1 \ : 1.78 \ : \ 35.7 \ : 285 \tag{134}$$

and the limit for the angle Θ_2

$$\Theta_2 < \sqrt{\frac{m_c}{m_t}} = 0.33 \tag{135}$$

For the assumptions given above Fritzsch obtained the following boundary

$$\sin\Theta_3 < 0.09 \qquad (\Theta_3 < 5^\circ) \tag{136}$$

However there is no agreement between descriptions of the quark masses ratios. The other authors [23] give smaller difference between quark masses

$$m_u \ : \ m_d \ : \ m_s \ : \ m_c = 1 \ : 1.1 \ : \ 6.4 \ : 23.6 \tag{137}$$

Thus for the ratio (137) we get the following limit for the angle Θ_3

$$\sin\Theta_3 < 0.163 \tag{138}$$

The value of the angle Θ_3 (130) is consistent with the boundary (138). The value (130) is close to the value given by Białas [24] and consistent with results in [25, 26] as well as the experimental boundary:

$$\sin\Theta_3 < 0.42 \qquad [24] \tag{139}$$

$$|\sin\Theta_3| = 0.28 \begin{cases} +0.21 \\ -0.28 \end{cases} \qquad [19] \tag{140}$$

Let us consider the relation between the angle Θ_2 and the phase parameter δ. From (125)–(127) we obtain

$$\beta + A = \frac{K - \frac{\pi}{2}}{K + B} \tag{141}$$

or equivalently

$$\frac{K + \frac{\pi}{2}}{K + B} - \frac{\gamma}{s_1^2} = s_2^2 \left(1 + \gamma \left(1 - \frac{2}{s_1^2}\right)\right) + A \tag{142}$$

denoting

$$\xi = \frac{\left(K + \frac{\pi}{2}\right) - \frac{\pi}{s_1^2}}{K + B} \tag{143}$$

$$\eta = 1 + \frac{\pi}{K + B} \tag{144}$$

$$\rho = \frac{1}{2} \quad \cos\Theta_1 \quad \sin 2\Theta_3 \tag{145}$$

we get

$$\cos\delta = \frac{\xi - s_2^2 \ \eta}{\rho \ \ \sin 2\Theta_2} \tag{146}$$

It is worth noting that if we take the constraint on the Cabibbo angle Θ_1 from the four- quark level [7, 16], which is the same as given by Eq. (107), the parameter ξ (143) will be exactly equal to zero, hence we get

$$\cos\ \delta = -\frac{\eta}{2\rho} \quad \tan\ \Theta_2 \tag{147}$$

Because $|cos\ \delta| \leq 1$, so from (147)

$$|\Theta_2| < \left| \arctan\ \frac{2\rho}{\eta} \right| \tag{148}$$

and we get also a boundary on the angle Θ_2

$$\sin\Theta_2 < 0.265 \qquad (\Theta_2 < 15.4^\circ) \tag{149}$$

The value (149) is in a good agreement with the results given by Fritzsch [15], Białas [24], Shrock, Treiman, Wang [22], Barger, Long, Pakvasa [25] and experimental limits [27], respectively:

$$9^\circ < \Theta_2 < 19^\circ \tag{150}$$

$$\sin\Theta_2 = 0.23 \tag{151}$$

$$|\sin\Theta_2| < 0.25 \qquad (m_t = 15\ \ \text{GeV}) \tag{152}$$

$$\sin\Theta_2 < 0.5 \qquad (m_t = 30\ \ \text{GeV}) \tag{153}$$

The Eq. (146) can be written as follows

$$x^2\ \ (\eta^2 + 4\ \rho^2\ \ \cos^2\delta) - 2x\left(\xi\eta\ \ + 2\rho^2 \cos^2\delta\right) + \xi^2 = 0 \tag{154}$$

where

$$x = \sin^2\Theta_2 \tag{155}$$

To get a real value of the angle Θ_2 the determinant of the square Eq. (154) cannot be negative, so

$$16\rho^2 \cos^2\delta\left(\xi\eta + \rho^2\cos^2\delta - \xi^2\right) \geq 0 \tag{156}$$

hence

$$1 \geq \cos^2\delta \geq \frac{\xi(\xi - \eta)}{\rho^2} \tag{157}$$

For (129, 130) we get

$$\eta = 0.9647 \qquad \rho = 0.1309 \tag{158}$$

If the parameter ξ. which can be identified with a change of the Cabibbo angle description by a transition to the higher symmetries, is slightly less than zero, the phase parameter δ will be bounded ($|\delta|$ should be nearly zero, as the Cabibbo angle description should not change strongly by a transition to higher symmetries, on the other hand the Eq. (157) gives a boundary on the parameter

$$\xi > -0.0175 \tag{159}$$

From (147)

$$\operatorname{sign}(\cos\delta) = -\operatorname{sign}(\tan\Theta_2) \tag{160}$$

so, for the angle Θ_2 lying in the first quadrant, it follows $\frac{\pi}{2} < \delta < \pi$ and from (159) there is a lower limit for the phase δ. For an input given by the Eqs. (129, 130, 132) we get

$$\xi = 0.002 \tag{161}$$

so there is no boundary on δ, since sign ξ = + 1. Let us notice, that a small change of the f_K can change the sign of the parameter ξ. Following Fuchs [28], in a chiral perturbation theory at the $SU_3 * SU_3$ level

$$\frac{f_K}{f_\pi} = 1 + \frac{3\left(m_K^2 - m_\pi^2\right)}{64\pi^2 f_\pi^2} \ln\frac{\Lambda}{4\mu^2} + O(\varepsilon) \tag{162}$$

where μ^2 is the average meson squared mass and Λ is a cut-off parameter, which is estimated to be near $4m_N^2$, it implies

$$\frac{f_K}{f_\pi} = 1.15 \tag{163}$$

$$\xi = -0.00179 \tag{164}$$

$$\cos^2\delta > 0.1 \tag{165}$$

Taking into account (160) we obtain

$$109^\circ < \delta < 180^\circ \tag{166}$$

Since f_K is treated as a variable and can depend on the energy scale via Λ parameter and the symmetry breaking parameters ε, the boundary of the phase due to the Eqs. (143, 157) can be expected. For sign $(\cos\delta) = -1$ there is a lower limit of the angle Θ_2 also. A variant $\cos\delta > 0$ is allowed but the angle Θ_2 corresponding to this variant is too severely limited and it is not consistent with the experimental data [27].

We have shown that the weak mixing angles at the six-quark level can be estimated in terms of the masses of pseudo-scalar mesons. The calculation of mixing angles is possible by using the hierarchical symmetry breaking leading to a quark masses generation. A number of independent mixing angles that can be calculated on the ground of the given above model is equal to a number of degrees of freedom connected with the symmetry breaking and the quarks mixing in the fixed electric charge subspace (let us notice that in the variant B after the rotation around the 21st axis and next around the 7th one, even the exact $SU_2 * SU_2$ symmetry did not remain; however the angle Θ_2 connected with the mixing in the positive electric charge subspace could be calculated). An assumption that in the hierarchical symmetry breaking the flavor does not have to be conserved on each stage of the symmetry breaking, while it is conserved in the broken symmetry taken as a whole, has allowed the author to introduce to the broken Hamiltonian density a phase angle responsible for CP-non-conservation. The experimental value of the Cabibbo angle treated as an input has allowed the author to calculate the angle Θ_3 and to find the relation connecting the angle Θ_2 and the phase parameter δ. Limits of trigonometric functions values imply boundaries on the angle Θ_2 and the phase δ. The kaon decay constant is a sensitive parameter, which can introduce CP-non-conservation to the chiral perturbation theory. Boundaries for the angle Θ_2 and the phase δ vs. f_K can be also found.

4. The standard six-quark model with a hierarchical symmetry breaking

The simultaneous mixing of quarks in both negative and positive electric charge sub-spaces is considered. Quark mixing in each space is described by the Kobayashi-Maskawa matrix. In order to get a right number of independent mixing parameters only one angle θ_7 common for both sub-spaces has been adjusted. Since the electromagnetic mass splitting of u and d quarks has been taken into account the real K-M mixing angles can be calculated explicitly. As an input only meson masses and f_x factors (treated as factors in matrix elements between one meson state and vacuum according to PCAC) are needed. Physical quark mixing is realized for maximal allowed symmetry breaking and it corresponds to vanishing of θ_7, which implies that only quark mixings with mass generation are permitted. Bounds of the phase δ have been also found.

The Kobayashi-Maskawa mixing matrix (69) is usually considered to mix quarks in the negative electric charge subspace. It can be written also as (70) and it can mix quarks either in the negative or in the positive electric charge subspace. A simultaneous mixing in both spaces [13] was also taken onto account.

Let us assume that quarks are mixed in both sub-spaces simultaneously, then the charged weak current is expressed as follows

$$J_\mu = (\bar{u}, \bar{c}, \bar{t})\ \gamma_\mu\ (1-\gamma_5)\ U_+\ U_- \begin{pmatrix} d \\ s \\ b \end{pmatrix} \tag{167}$$

where

$$U_+ = \begin{pmatrix} 1 & 0 & 0 \\ 0 & c_2 & s_2 \\ 0 & -s_2 & c_2 \end{pmatrix} \begin{pmatrix} c_5 & s_5 & 0 \\ -s_5 & c_5 & 0 \\ 0 & 0 & 1 \end{pmatrix} \begin{pmatrix} 1 & 0 & 0 \\ 0 & 1 & 0 \\ 0 & 0 & e^{i\delta_2} \end{pmatrix} \begin{pmatrix} 1 & 0 & 0 \\ 0 & c_6 & s_6 \\ 0 & -s_6 & c_6 \end{pmatrix} \tag{168}$$

$$U_- = \begin{pmatrix} 1 & 0 & 0 \\ 0 & c_7 & s_7 \\ 0 & -s_7 & c_7 \end{pmatrix} \begin{pmatrix} 1 & 0 & 0 \\ 0 & 1 & 0 \\ 0 & 0 & e^{i\delta_1} \end{pmatrix} \begin{pmatrix} c_1 & s_1 & 0 \\ -s_1 & c_1 & 0 \\ 0 & 0 & 1 \end{pmatrix} \begin{pmatrix} 1 & 0 & 0 \\ 0 & c_3 & s_3 \\ 0 & -s_3 & c_3 \end{pmatrix} \tag{169}$$

The matrices U_+ and U_- mix quarks in spaces with charges +2/3 and − 1/3 respectively. The Kobayashi-Maskawa mixing matrix is parametrized by four independent parameters only, while in the product of the matrices (168) and (169) in the current (167) there are eight mixing angles. In order to get the effective mixing matrix in the K-M form with the right number of independent mixing parameters, we must adjust the angles in such a way as to get the effective matrix with only four independent angles. We shall demand the following elements U_{11}, U_{12}, U_{13}, U_{21}, U_{31} of the effective matrix to be real and the complex phase to exist in the elements U_{22}, U_{23}, U_{32}, U_{33} only, as in the original K-M matrix. The only solution is

$$\theta_6 = -\theta_7 \tag{170}$$

Hence in the matrix $U_+ U_-$ there will be effectively only four parameters: θ_2, $\theta_C = \theta_1 + \theta_5$, θ_3 and $\delta = \delta_1 + \delta_2$. The current (167) can be expressed as follows

$$J_\mu = R_2 \ R_1 \ J_\mu(0) \ R_1^{-1} \ R_2^{-1} \tag{171}$$

where

$$J_\mu(0) = (\overline{u}, \overline{c}, \overline{t}) \ \gamma_\mu \ (1 - \gamma_5) \ I \begin{pmatrix} d \\ s \\ b \end{pmatrix} \tag{172}$$

$$R_1 = e^{-2i\theta_3 Q^{21}} e^{-2i\theta_1 Q^7} \ e^{-iX\delta_1} \ e^{-2i\theta_7 Q^{21}} \tag{173}$$

$$R_2 = e^{-2i\theta_2 Q^{32}} e^{-iY\delta_2} \ e^{-2i\theta_5 Q^{10}} \ e^{-2i\theta_7 Q^{32}} \tag{174}$$

where X = (79) and $Y = \frac{\sqrt{15}}{3} Q^{35}$. Q^k is the $6 * 6$ matrix representation of the k-th generator of SU_6 group. In variants A (72) and B (73), because of quark mixing in (d, s, b) sector only, the electromagnetic mass splitting of u and d quarks was neglected. For the simultaneous mixing in both (d, s, b) and (u, c, t) sectors the calculation of the angles θ_i explicitly is not possible (see below formulas (187) and (188). The Hamiltonian density breaking the chiral $SU_6 * SU_6$ symmetry is given as follows

$$H_0 = \sum_{j=1}^{6} c_{j^2-1} \ u^{j^2-1} \tag{175}$$

where c_i are the symmetry breaking parameters, u^i- the scalar components of the $(\overline{6}, 6) + (6, \overline{6})$ of the chiral $SU_6 * SU_6$ group. From the GMOR model we obtain the following relations for masses of pseudo-scalar mesons for $SU_6 * SU_6$ symmetry:

$$\pi = m_\pi^2\ f_\pi^2 = Z\left(\frac{c_0}{\sqrt{3}} + \frac{c_8}{\sqrt{3}} + \frac{c_{15}}{\sqrt{6}} + \frac{c_{24}}{\sqrt{10}} + \frac{c_{35}}{\sqrt{15}}\right) \tag{176}$$

$$K^+ = m_{K^+}^2\ f_{K^+}^2 = Z\left(\frac{c_0}{\sqrt{3}} + \frac{c_3}{2} - \frac{c_8}{2\sqrt{3}} + \frac{c_{15}}{\sqrt{6}} + \frac{c_{24}}{\sqrt{10}} + \frac{c_{35}}{\sqrt{15}}\right)$$

$$K^0 = m_{K^0}^2\ f_{K^0}^2 = Z\left(\frac{c_0}{\sqrt{3}} - \frac{c_3}{2} - \frac{c_8}{2\sqrt{3}} + \frac{c_{15}}{\sqrt{6}} + \frac{c_{24}}{\sqrt{10}} + \frac{c_{35}}{\sqrt{15}}\right)$$

$$D^+ = m_{D^+}^2\ f_{D^+}^2 = Z\left(\frac{c_0}{\sqrt{3}} - \frac{c_3}{2} + \frac{c_8}{2\sqrt{3}} - \frac{c_{15}}{\sqrt{6}} + \frac{c_{24}}{\sqrt{10}} + \frac{c_{35}}{\sqrt{15}}\right)$$

$$D^0 = m_{D^0}^2\ f_{D^0}^2 = Z\left(\frac{c_0}{\sqrt{3}} - \frac{c_3}{2} + \frac{c_8}{2\sqrt{3}} - \frac{c_{15}}{\sqrt{6}} + \frac{c_{24}}{\sqrt{10}} + \frac{c_{35}}{\sqrt{15}}\right)$$

$$B^+ = m_{B^+}^2\ f_{B^+}^2 = Z\left(\frac{c_0}{\sqrt{3}} + \frac{c_3}{2} + \frac{c_8}{2\sqrt{3}} + \frac{c_{15}}{2\sqrt{6}} - \frac{3c_{24}}{2\sqrt{10}} + \frac{c_{35}}{\sqrt{15}}\right)$$

$$T^+ = m_{T^+}^2\ f_{T^+}^2 = Z\left(\frac{c_0}{\sqrt{3}} - \frac{c_3}{2} + \frac{c_8}{2\sqrt{3}} + \frac{c_{15}}{2\sqrt{6}} + \frac{c_{24}}{2\sqrt{10}} - \frac{2c_{35}}{\sqrt{15}}\right)$$

In a model with hierarchical symmetry breaking the highest exact symmetry, which can be assumed, is the $SU_4 * SU_4$ one. At least one quark in each sector must be massive. Following the procedure described in [29] the Hamiltonian density breaking the chiral $SU_6 * SU_6$ symmetry will be rotated in the opposite direction by comparison with the rotation of the weak charged current.

$$H_{SB} = R_{21}\ R_{11}\ H_E\ R_{11}^{-1}\ R_{21}^{-1} \tag{177}$$

where

$$R_{11} = e^{-2i\theta_7 Q^{21}}\, e^{-iX\delta_1}\, e^{-2i\theta_1 Q^7}\ e^{-2i\theta_3 Q^{21}} \tag{178}$$

$$R_{21} = e^{-2i\theta_7 Q^{32}}\, e^{-iY\delta_2}\ e^{2i\theta_5 Q^{10}}\ e^{2i\theta_2 Q^{32}}$$

The exact $SU_4 * SU_4$ symmetry implies that

$$c_3 = c_8 = c_{15} = \sqrt{5}c_0 + c_{35} = 0 \tag{179}$$

The $SU_4 * SU_4$ invariant Hamiltonian density is given as

$$H_E = P\overline{q}_6 q_6 - V\overline{q}_5 q_5 \tag{180}$$

where

$$P = \sqrt{12}c_0 + V \qquad V = \frac{5}{\sqrt{10}}c_{24} \tag{181}$$

We shall assume that in the model with hierarchical symmetry breaking the flavor will not be conserved in the intermediate stages of the symmetry breaking, but it will be conserved in the broken symmetry taken as a whole. The symmetry breaking Hamiltonian density retaining onlyÂ· the flavor-conserving part is given as follows

$$\begin{aligned} H_{(\Delta F=0)} = {} & \overline{q}_6 q_6 P(\lambda - M) - \overline{q}_5 q_5 V(\alpha - A) + \overline{q}_4 q_4 P(\rho + M) - \overline{q}_3 q_3 V(\beta + A) - \overline{q}_2 q_2 V\gamma \\ & + \overline{q}_1 q_1 P\tau \end{aligned} \tag{182}$$

where

$$\begin{aligned} \alpha &= c_1^2 s_3^2 s_7^2 + c_3^2 c_7^2 \\ \beta &= c_1^2 s_3^2 c_7^2 + c_3^2 s_7^2 \\ \gamma &= s_1^2 s_3^2 \\ A &= \frac{1}{2} \sin 2\theta_7 \sin 2\theta_3 c_1 \cos \delta_1 \end{aligned} \tag{183}$$

$$\begin{aligned} \lambda &= s_2^2 c_5^2 s_7^2 + c_2^2 c_7^2 \\ \rho &= s_2^2 c_5^2 c_7^2 + c_2^2 s_7^2 \\ \tau &= s_2^2 s_5^2 \\ M &= -\frac{1}{2} \sin 2\theta_7 \sin 2\theta_2 c_5 \cos \delta_2 \end{aligned} \tag{184}$$

The broken Hamiltonian density (182) can be expressed as a function of operators u^k (k = 0, 3, 8, 15, 24, 35). The coefficients of the operators u^k are as follows

$$\begin{aligned} c_0' &= c_0 \\ c_3' &= \frac{1}{2}(P\tau + V\gamma) \\ c_8' &= \frac{1}{2\sqrt{3}}(P\tau + V(2\beta' - \gamma)) \\ c_{15}' &= \frac{1}{2\sqrt{6}}(P(\tau - 3\rho') - V(\beta' + \gamma)) \\ c_{24}' &= \frac{1}{2\sqrt{10}}(P(\tau + \rho') - V(5\beta' + 5\gamma - 4)) \\ c_{35}' &= \frac{1}{2\sqrt{15}}(6P(\tau + \rho') - 5P - V) \end{aligned} \tag{185}$$

where

$$\beta' = \beta + A \qquad \rho' = \rho + M \tag{186}$$

After symmetry breaking the pseudo-scalar masses (176) will be described as functions of the coefficients c_i' [16].

$$\begin{aligned} &\pi = \frac{Z}{2}(P\tau - V\gamma) \\ &K^+ = \frac{Z}{2}(P\tau - V\beta') && K^0 = -\frac{Z}{2}V(\beta' + \gamma) \\ &D^0 = \frac{Z}{2}P(\rho' + \tau) && D^+ = \frac{Z}{2}(P\rho' - V\gamma) \\ &B^+ = \frac{Z}{2}(P\tau + V(\beta' + \gamma - 1)) && T^+ = \frac{Z}{2}(P(1 - \tau - \rho') - V\gamma) \end{aligned} \tag{187}$$

From (183), (184), (186) and (187) we get

$$\beta' = \frac{K^0 + K^+ - \pi}{2B^+ + 3K^0 - K^+ - \pi} \qquad \gamma = \frac{K^0 - K^+ + \pi}{2B^+ + 3K^0 - K^+ - \pi}$$
$$\rho' = \frac{D^0 + D^+ - \pi}{2T^+ + 3D^0 - D^+ - \pi} \qquad \tau = \frac{D^0 - D^+ + \pi}{2T^+ + 3D^0 - D^+ - \pi} \tag{188}$$

(contrary to the case of mixing in (d, s, b) sector only (variant A in [29] the electromagnetic mass splitting of u and d quarks cannot be neglected; if we put arbitrarily $c_3 = 0$ in Eq. (175) as in variants A and B in [29], the parameters $\gamma, \beta'\tau, \rho'$ could not be calculated separately. We would obtain only three nonlinear relations connecting these parameters with meson masses). Since

$$\alpha + \beta + \gamma = \lambda + \rho + \tau = 1 \tag{189}$$

putting (183, 184, 186) to (188) and eliminating θ_2 and θ_3 from the obtained set of four equations we get

$$f_1(\theta_1, \delta_1) = \tan\theta_7 = -f_5(\theta_5, \delta_2) \tag{190}$$

where

$$f_1(\theta_1, \delta_1) = \frac{B_1 \mp \sqrt{B_1^2 - A_1 C_1}}{A_1} \tag{191}$$

$$A_1 = s_1^2(1 - \beta') - \gamma \qquad B_1 = \sqrt{\gamma}\sqrt{s_1^2 - \gamma c_1 \cos\delta_1} \qquad C_1 = \gamma - s_1^2(\beta' + \gamma) \tag{192}$$

$$f_5(\theta_5, \delta_2) = \frac{B_5 \mp \sqrt{B_5^2 - A_5 C_5}}{A_5} \tag{193}$$

$$A_5 = s_5^2(1 - \rho') - \tau \qquad B_5 = \sqrt{\tau}\sqrt{s_5^2 - \tau c_5 \cos\delta_2} \qquad C_5 = \tau - s_5^2(\rho' + \tau) \tag{194}$$

We considered in [16] the simultaneous mixing in (d, s) and (u, c) sectors in the $SU_4 * SU_4$ symmetry. The mixing angles Θ and ϕ could not be calculated separately, however the nonlinear formula connecting both angles and pseudo-scalar masses was found

$$2\pi + 2(K + D)\sin^2\Theta \sin^2\phi = (2K + \pi)\sin^2\Theta + (2D + \pi)\sin^2\phi \tag{195}$$

A numerical calculation showed that there is an extremum (a maximum) of the function (66) with condition (195) for the angles $\Theta_m + \phi_m$ very close to the experimentally measured Cabibbo angle. This fact suggests that the symmetry breaking is realized in the maximal allowed case, so the effective angle of mixing would correspond to the maximum of function (66). As in [16] we shall look for the extremum of the function

$$f(\theta_1, \theta_5) = \sin(\theta_1 + \theta_5) \tag{196}$$

with condition (190). The following set of equations must be obeyed

$$f_1(\theta_1, \delta_1) + f_5(\theta_5, \delta_2) = 0 \qquad \frac{\partial f_1(\theta_1, \delta_1)}{\partial \delta_1} = 0 \tag{197}$$

$$\frac{\partial f_1(\theta_1, \delta_1)}{\partial \theta_1} - \frac{\partial f_5(\theta_5, \delta_2)}{\partial \theta_5} = 0 \qquad \frac{\partial f_5(\theta_5, \delta_2)}{\partial \delta_2} = 0$$

From (197) we get

$$C_1 = 0 \qquad C_5 = 0 \tag{198}$$

respectively, which implies that the separation constant

$$\tan\, \theta_7 = 0 \tag{199}$$

This means that the maximal allowed symmetry breaking occurs only for independent mixing of quarks in both sectors.

Let us consider the action of the operators R_{11} and R_{21} on quarks. The operator R_{11} mixes quarks in the negative electric charge subspace in the following sequence: (s-b)(θ_3), (d-s)(θ_1), a phase rotation (δ_1), (s-b)(θ_7), however the operator R_{21} mixes quarks as follows: (c-t) (θ_2), (u-c)(θ_5). a phase rotation (δ_2), (c-t)(θ_7). By the exact $SU_4 * SU_4$ symmetry only b and t quarks are massive. After the symmetry breaking a massless quark can become massive if it mixes with the other massive one. By the mixing in the sector with the charge - 1/3 the quark s has become massive in the first stage of the hierarchical symmetry breaking, after mixing with the quark b (the rotation on the angle θ_3 generated by the operator Q^{21}), the quark d has become massive in the second stage after mixing with the already massive quarks (the rotation on the angle θ_1). The next rotation by the angle θ_7, and mixing of s and b quarks are not connected with the symmetry breaking, because the mixing quarks have been already massive. There is analogical situation in the sector with the charge +2/3. The c and u quarks have become massive due to the hierarchical symmetry breaking (rotations on angles θ_2 and θ_5, respectively), however the rotation by the angle θ_7 and mixing of c and t quarks are also not connected with the symmetry breaking. Thus, from (199) it results that the physical quark mixing is realized only in the symmetry breaking with the quark masses generation. Putting (199) to (183) and (184) and comparing with (188) we get

$$\sin^2\theta_1 = \frac{K^0 - K^+ + \pi}{2K^0} \qquad \sin^2\theta_5 = \frac{D^0 - D^+ + \pi}{2D^0} \tag{200}$$

$$\sin^2\theta_3 = \frac{2K^0}{2B^+ + 3K^0 - K^+ - \pi} \qquad \sin^2\theta_2 = \frac{2D^0}{2T^+ + 3D^0 - D^+ - \pi}$$

so $\theta_C = \theta_1 + \theta_5$ depends on the parameters of mesons belonging only to the SU_4 multiplet. Let us notice that in comparison to the variant A in [29], taking into account the quark mixing in the (u, c, t) sector allowed the author to calculate the Cabibbo angle from the model and the angles θ_2 and θ_3. Let us compare the value of the calculated angle $\theta_C = \theta_1 + \theta_5$ realized for the maximal symmetry breaking with the experimentally measured Cabibbo angle value [22].

$$\cos\theta = 0.9737 \pm 0.0025 \tag{201}$$

The well known values of meson masses were taken from [30]. $f_\pi, f_{K^+}, \ldots$ were assumed as the factors in the matrix elements between one meson state and the vacuum according to PCAC, so for meson multiplets with the isospin 1/2 the factors for charged and neutral mesons are the same [31]. There exist many conjectures concerning the values of f_x. They widely differ in magnitude, depending on the particular approach to the estimation of the matrix element $< 0|v_x|x >$ and so far they have no reliable experimental support. Only in the case of f_K there is a fair consensus that the value is around 1.28 [12, 13, 31, 32]. For a calculation we took as f_D for comparison's sake values significantly different

$$f_D = 0.974 \quad [10] \qquad\qquad f_D = 0.65 \quad [29] \tag{202}$$

which gives

$$\cos\,(\theta_1 + \theta_5) = 0.9799 \qquad\qquad \cos\,(\theta_1 + \theta_5) = 0.9709 \tag{203}$$

respectively, very close to the experimental value (201), as in the case of the $SU_4 * SU_4$ broken symmetry [16]. It seems to us that such a well agreement in both $SU_4 * SU_4$ and $SU_6 * SU_6$ symmetries is not accidental and the symmetry breaking is indeed realized for the maximal allowed case.

Putting (198) to (197) we find the relation connecting both phase parameters

$$\cos\,\delta_2 = \xi\,\cos\,\delta_1 \tag{204}$$

where

$$\xi = \sqrt{\frac{\gamma\rho'\,(1-\gamma-\beta)\,(\rho'+\tau)}{\tau\beta'\,(1-\tau-\rho')\,(\beta'+\gamma)}} \tag{205}$$

The effective phase parameter $\delta = \delta_1 + \delta_2$ is bounded for $\xi \neq 1$. Indeed, the Eq. (204) has solution only for

$$|\delta| \begin{cases} > \arccos\,\dfrac{1}{\xi} & \text{if } (\xi > 1) \\ > \arccos\,(\xi) & \text{if } (\xi < 1) \end{cases} \tag{206}$$

It is worth noticing that even for $\xi \to \infty$ or $\xi \to 0$ the second and third quadrant for δ is still allowed.

Appendix

From the Gell-Mann Oakes Renner model for $SU_6 * SU_6$ symmetry we obtain the following relation for masses of pseudo-scalar mesons

$$m_a^2 f_a^2 \delta^{ab} + \int \frac{dq^2}{q^2}\,\rho^{ab} = i < 0\left|\left[\overline{Q}^a, \overline{D}^b\right|0 > = \sum_{i=1}^{6}\left(\sum_{j=1}^{6} c_{j^2-1} d_{i^2-1,a,c} d_{j^1-1,b,c}\right) < u^{i^2-1} >_0\right. \tag{207}$$

where

$$\rho^{ab} = (2\pi)^3 \sum_{n\neq a} \delta^4\,(p_n - q) < 0|\overline{D}^a|n> < n|\overline{D}^b|0> \tag{208}$$

d_{abc} symmetric constants of the SU_6 group, $< u^i >_0$ - vacuum expectation value of the operator u^i, $Q^i \pm \overline{Q}^i = \int d3^x\, V_0^\alpha(x) \pm \int d^3x A_0^\alpha(x)$ - the generators of the $SU_6 * SU_6$ group

$$D^a = \partial^\mu V_\mu^a(x) \qquad \overline{D}^a = \partial^\mu A_\mu^a(x) \qquad [12] \tag{209}$$

Because the vacuum expectation values of operators u^i: i = 3, 8, 15, 24, 35 and the spectral density ρ^{ab} are proportion to the squared parameters Of symmetry breaking, they were neglected. Approximately we obtain

$$m_a^2 f_a^2 = \frac{1}{\sqrt{3}} \left(\sum_{j=1}^{6} c_{j^2-1} d_{j^2-1,a,a} \right) < u_0 >_0 \tag{210}$$

Because the symmetric constants of SU_6 group: $d_{113} = d_{223} = d_{333} = 0$ the masses of neutral and charged pions are not differentiated, however there is the electromagnetic mass splitting of the other meson multiplets (see Eq. (176)).

The experimental data [30] gives

$$\Delta m_K = m_{K^0} - m_{K^+} = 4.003 \ MeV \qquad \Delta m_D = m_{D^0} - m_{D^+} = -5.3 \ MeV \tag{211}$$

so

$$sign \ \ \Delta m_K = -sign \ \ \Delta m_D \tag{212}$$

Let us notice that from (17613) we get

$$sign \ \ (K^0 - K^+) = -sign \ \ (D^0 - D^+) \tag{213}$$

so the direction of the electromagnetic mass splitting by the factor c_3 responsible for this effect is consistent with the experimental data. On the other hand

$$sign \ \ \Delta m_\pi = m_{\pi^0} - m_{\pi^+} = -4.603 MeV \tag{214}$$

so the electromagnetic mass splitting of pions is of the same order as kaons or D mesons. It suggests that the neglected terms in approximate formula (207) are of the order of the factor c_3. This means that such an approximation does not generate error greater than the electromagnetic ma s splitting of pion in meson masses description.

Author details

Zbigniew Piotr Szadkowski
Faculty of Physics and Applied Informatics, University of Łódź,
90-236 Pomorska 149, Łódź, Poland

*Address all correspondence to: zbigniew.szadkowski@fis.uni.lodz.pl

References

[1] E.S. Abers, B.W. Lee, Physics Reports 9, Issue 1 (1973)

[2] S. Weinberg, Phys. Rev. Lett. 19, 1264 (1967)

[3] A. Salam, Rev. of Modern Phys. 52, 539 (1980)

[4] R. J. Oakes, Phys. Lett. 298, 683 (1969).

[5] A. Ebrahim, Lett. Nuovo Cimento 19, 225 (1977).

[6] A. Ebrahim, Lett. Nuovo Cimento 19, 437 (1977).

[7] Z. Szadkowski, Acta Phys. Pol. B13, 247 (1982).

[8] A. Ebrahim, Phys. Lett. 698, 229 (1977).

[9] M. Gell-Mann, R. J. Oakes, B. Renner, Phys. Rev. 175, 2195 (1968).

[10] N. Cabibbo, Phys. Rev. Lett. 10, 531 (1963).

[11] S. L. Glashow, J. Iliopoulos, L. Maiani, Phys. Rev. D2, 1285 (1970).

[12] V. de Alfaro, S. Fubini, G. Furlan, C. Rossetti, Currents in Hadron Physics, North Holland Publishing Company 1973, p. 509–514.

[13] K. P. Das, N. G. Deshpande, Phys. Rev. D19, 3387 (1979).

[14] V. P. Gautam, R. Bagchi, Acta Phys. Pol. B9, 1017 (1978).

[15] H. Fritzsch, Nucl. Phys. BISS, 189 (1979).

[16] Z. Szadkowski, Acta Phys. Pol. Bl4, 23 (1983).

[17] Y. Fujimoto, Lett. Nuovo Cimento 29, 283 (1980).

[18] R. Mignani, Lett. Nuovo Cimento 28, 529 (1980).

[19] M. Kobayashi, K. Maskawa, Prog. Theor. Phys. 49, 652 (1973).

[20] N. Cabibbo, L. Maiani, Phys. Lett. 28B, 131 (1968).

[21] R. Gatto, G. Sartori, M. Tonin, Phys. Lett. 28B, 128 (1968).

[22] R. E. Shrock, S. B. Treiman, L. L. Wang, Phys. Rev. Lett. 12, 1589 (1979).

[23] B. Bagchi, V. P. Gautam, A. Nandy, Phys. Rev. 019, 3380 (1979).

[24] A. Białas, Acta Phys. Pol. B12, 897 (1981).

[25] V. Barger, W. F. Long, S. Pakvasa, Phys. Rev. Lett. 42, 1585 (1979).

[26] E. Ma, Phys. Lett. 96B, 115 (1980).

[27] K. Kleinknecht, B. Renk, Z. Phys. C-Particles and Fields 16, 7 (1983).

[28] N. H. Fuchs, Lett. Nuovo Cimento 27, 21 (1980).

[29] Z. Szadkowski, Acta Phys. Pol. B16, 1067 (1985).

[30] Particle Data Group, Phys. Lett. 1118, (1982).

[31] S. L. Adler, R. F. Dashen, Current Algebras and Applications to Particle Physics, W. A. Benjamin Inc. N. Y.-Amsterdam 1968, p. 125–142.

[32] S. Fajfer, Z. Stipcevic, Lett. Nuovo Cimento 29, 207 (1980).

Chapter 3

The Inter-Nucleon Up-to-Down Quark Bond and Its Implications for Nuclear Binding

Nancy Lynn Bowen

Abstract

This paper describes an interesting and potentially significant phenomenon regarding the properties of up and down quarks within the nucleus, specifically how the possible internucleon bonding of these quarks may affect the bonding energy of the nuclear force. A very simple calculation is used, which involves a bond between two internucleon up and down quarks. This simple calculation does not specify the shape or structure for the nucleus, rather this calculation only examines the energy of all possible internucleon up-to-down bonds that may be formed within a quantum nucleus. A comparison of this calculated binding energy is made to the experimental binding energy with remarkably good results. The potential significance and implications of this noteworthy finding are discussed.

Keywords: nuclear binding energy, nuclear force, nuclear bond, quarks, up quark, down quark, internucleon bond, quantum

1. Introduction

The nuclear force is defined as the force which binds the protons and neutrons together within a nucleus. One of the currently accepted models of the nuclear force is the liquid drop model [1]. This model of the nuclear force uses the Weizsäcker formula to predict the binding energies of nuclides. The Weizsäcker formula is a curve-fitting formula that uses five parameters, plus one conditional logic statement, in order to achieve its results [2]. These parameters are selected to empirically curve-fit an equation to match the experimental data. The liquid drop model is considered to be a "semi-classical" model of the nuclear force, rather than a quantum model [3].

Another currently accepted model of the nuclear force is the shell model, which uses magic numbers to explain certain nuclear behavior. The nuclear shell model is similar to the electronic shell model, which describes the electrons orbiting around an atom. However, the nuclear shell model does not predict the nuclear binding energy, rather the shell model defers back to the Weizsäcker formula for binding energy calculations.

A third currently accepted model of the nuclear force is the residual chromodynamic force (RCDF) model, also known as the residual strong force model.

Before describing this residual chromodynamic force, it is useful to mention a few specifics about quantum chromodynamics (QCD). Quantum chromodynamics postulates that the three valance quarks of protons and neutrons possess an attribute called "color charge." Historically, a contradiction of the quantum mechanical basis of nucleon properties with the Pauli Exclusion Principle led to the concept of the color charge for quarks [4]. The color charges of the quarks are considered to be either red, green, or blue. The words red, green, and blue are simply the names of the color charges and do not imply any type of physically visual hue for the quarks. Also, the term "charge", when referring specifically to the color charge, is not related to electric charge, which unfortunately can often be a point of confusion. Quantum chromodynamics states that a very strong bond is formed among the three color charges of the quarks inside the nucleon [5]. Both protons and neutrons have all three colors inside the nucleon.

The residual chromodynamic force model assumes that the chromodynamic force also has a weaker *residual* force outside of the nucleon. The RCDF model states that this residual force forms an internucleon bond, binding the nucleons together. The internucleon bond is formed by the residual chromodynamic force of the quarks outside of the nucleons. This is shown, in an illustrative representation, in **Figure 1**.

In **Figure 1**, the bold black line represents the chromodynamic force inside the nucleon, and the dotted gray line represents the residual chromodynamic force between two nucleons. (Note that quarks are considered to be point-like particles. Thus this drawing is not meant to be a scaled representation of the quarks, rather it is meant for illustrative purposes only.) The residual chromodynamic bond can be between any two quarks of different colors, such as between a red and blue, a green and red, or a blue and green. The residual chromodynamic bond can be between a neutron and a neutron, a proton and a neutron, or a proton and another proton.

While the RCDF model is considered to be the mechanism for nuclear bonding, the model is unable to duplicate the experimental binding energy curve. This inability of the RCDF model to reproduce this nuclear behavior is currently attributed to the extreme difficulty of modeling the multi-body interactions of the three color charges [6, 7]. This difficulty with the derivation of the nuclear binding forces from the residual chromodynamic force model is two-fold. First, each nucleon consists of three quarks, which means that a system of two nucleons is already a six-body problem. Second, because the chromodynamic force between quarks inside the nucleons has the feature of being very strong compared to the residual chromodynamic force outside the nucleons, this disproportionate ratio of strength makes a converging solution for the complicated mathematical calculations difficult to find. For nuclides with a small number of nucleons, the problem can be solved with

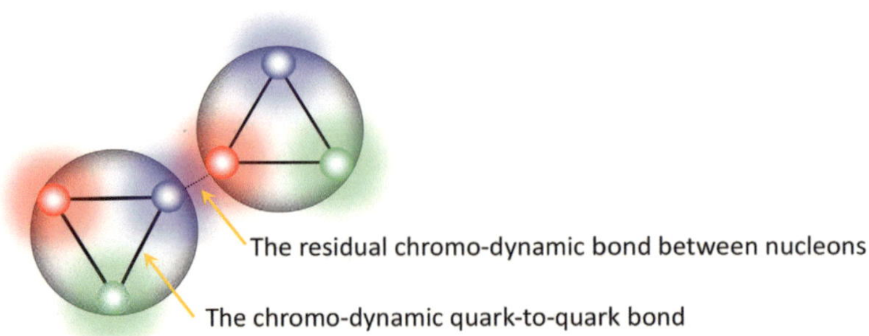

Figure 1.
An illustrative representation showing both the chromodynamic force and the residual chromodynamic force.

brute-force computing power by putting each of the quarks into a four-dimensional lattice of discrete points: three dimensions of space and one of time. This method is known as lattice quantum chromodynamics, or lattice QCD. This brute-force method for the computer calculations in lattice QCD iterates the position of each quark by assigning an x, y, z, and t position to it, calculating the resulting forces on each quark, allowing their position to change as a result of these forces, and then iterating this procedure until a resulting converging solution is found. If a converging solution is found, then these calculations are able to determine the binding energy of the nuclide in question. These computer calculations are done through extremely complex mathematical models, often using Monte-Carlo simulations [8]. Because of these computational difficulties, modeling the binding energies of only the smallest nuclides has been achieved.

However, such calculations are computationally expensive, requiring very large computers. Because of these complications, this modeling method is not normally used as a standard nuclear physics tool [7] .Thus, the RCDF model remains largely unverified when testing its binding energy predictions against experimental data.

2. Properties of up and down quarks

Besides having the attribute of color charge, there is also another attribute of quarks called flavor. From QCD theory, we know there are six different flavors of quarks: up, down, strange, charm, top, and bottom. Of these six different flavors, only two flavors are found in the stable matter of neutrons and protons: the up and down quarks [9]. (The terms of up and down do not imply any specific orientation with regard to spatial direction, and are simply the names of these types of quarks).

An up quark has an electric charge that is +2/3 the charge of a proton, and it also contains a positive magnetic moment. The up quark has a spin of ½ and a mass of about 0.3% of the proton. The color of an up quark can be either red, green, or blue. A down quark has an electric charge that is −1/3 the charge of a proton, and it contains a negative magnetic moment, which is anti-parallel to of the spin of the nuclide. The down quark has a spin of ½, and a mass of about 0.6% of the proton. The color of a down quark can be either red, green, or blue.

The magnetic moments of an up quark is estimated to be +1.85 and the magnetic moments of a down quark is estimated to be −0.97, both in units of nuclear magnetons. The electric charges of the proton and neutron are completely contained within the quarks. The proton is comprised of two up quarks and one down quark, giving it a net charge of one (2/3 + 2/3 -1/3 = 1). The neutron is comprised of one up quark and two down quarks, giving it a net charge of zero (2/3 -1/3 -1/3 = 0). **Figure 2** illustrates these properties.

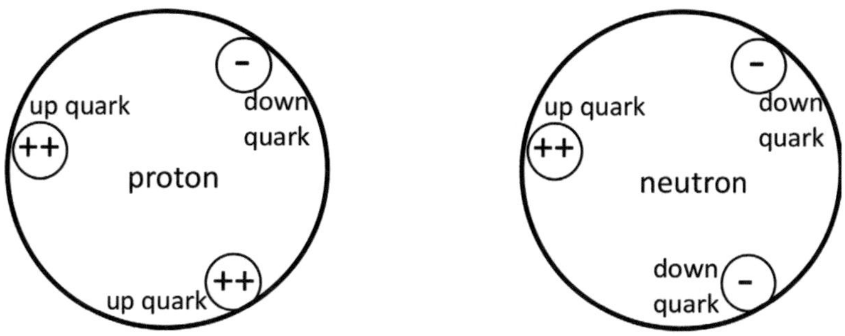

Figure 2.
An illustrative representation of the up and down quarks in a proton and neutron.

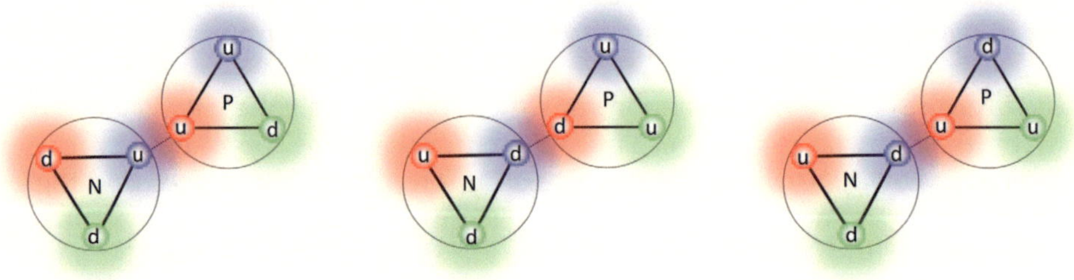

Figure 3.
Possible bonds in the RCDF model. A bond can be formed regardless of the flavor (up or down) of the quark.

The quarks inside of a proton and neutron have both attributes of flavor (up or down) and color (red, green, or blue). Thus, each quark inside of a proton or neutron is one of six types: up and red, up and green, up and blue, down and red, down and green, or down and blue [5]. Since both the neutron and the proton contain all three different colors, there is no difference between the proton and the neutron with regard to the attribute of color charges. The only difference in the quark characteristics between a proton or a neutron resides in the number of up and down quarks. Therefore, any bond between the different colors is also inherently a bond between some combination of the up and down quarks. Hence, the quantum assumptions that are made in the RCDF model about the possibility of an internucleon bond between the residual colors of quarks are also inherently applicable to the formation of an internucleon bond between up and down quarks.

Figure 3 shows three possible bonds, all of which are allowed in the RCDF model: a bond between two up quarks, between two down quarks, and between an up and a down quark. In the RCDF model, as long as the bond is between different colors of quarks, the up or down flavor of the quarks, is considered relatively unimportant.

Although it is considered relatively unimportant in the RCDF model, the up or down flavor of the quarks does indeed cause an energy difference among the three types of bonds that are illustrated in **Figure 3**. If there is an internucleon bond between two up quarks or between two down quarks, then the intrinsic electromagnetic force between these quarks is repulsive. Conversely, if there is an internucleon bond between an up quark and a down quark, then the intrinsic electromagnetic force between these quarks is attractive. Among the three types of bonds shown in **Figure 3**, this inherent difference in the electromagnetic energy may cause the two repulsive bonds to be less probable or less stable, producing a situation in which the up-to-down quark bond would be more prevalent in stable matter.

3. An additional constraint for internucleon quark-to-quark binding

As mentioned previously, the color charges of the quarks contained within a nucleon do not inherently distinguish between a neutron or proton; it is only the up and down attribute of the quarks that distinguish between the two types of stable nucleons. Thus, an examination of an internucleon bond being formed only between an up and a down quark is an appropriate possibility to explore. Specifically, this additional constraint is that not only must the internucleon quark-to-quark bond be between different colors, but also it must be between only an up and a down quark; specifically, it cannot be between two up quarks or two down quarks. If this quite reasonable constraint is made to the RCDF model, a quick calculation of the allowed bonds can be easily made. By using the currently accepted RCDF concept of the internucleon quark-to-quark bond, and applying this additional constraint, in which bonds are only formed between up and down quarks, an interesting and potentially significant set of data emerges.

For any given nuclide, the number of internucleon up-to-down quark pairs can be determined, based on how many up and down quarks each nuclide has. This calculation, as shown in Eq. (1), is made for each nuclide.

$$\begin{gathered} Number_{up\ quarks} = (Z \times 2) + (N \times 1) \\ Number_{down\ quarks} = (Z \times 1) + (N \times 2) \\ Number_{possible\ pairs} = the\ smaller\ of\ these\ two\ numbers \end{gathered} \tag{1}$$

For simplicity of this very quick and easy calculation, it is assumed that every bonded pair of up-to-down quarks has the same bonding energy. Thus, just for this simple calculation, the equation for the calculated binding energy (CBE) of a non-quantum nuclide is the number of internucleon up-to-down quark pairs times the binding energy per pair, as shown in Eq. (2).

$$CBE = (number\ of\ pairs) \times (binding\ energy\ per\ bonded\ pair) \tag{2}$$

For a representative sample of stable nuclides, this information is also shown in **Table 1**. For values of mass number with two stable nuclides, such as A = 40, both stable nuclides are shown. The following information is listed:

- The nuclide name
- The number of nucleons, A
- The number of protons, Z
- The number of neutrons, N
- The experimental binding energy (EBE) in units of MeV, as obtained from the nuclear tables in Ref. [10].
- The experimental binding energy per nucleon (EBE/A)
- The number of up quarks in the nuclide
- The number of down quarks in the nuclide
- The number or possible pairs between up and down quarks for the nuclide
- The classical (non-quantum) calculated binding energy (CBE) in MeV of the nuclide, for a fixed energy (6.000 MeV) per bond.

Figure 4 is a plot for this same a representative sample of nuclides, showing both the experimental binding energy per nucleon (EBE/A) and the non-quantum calculated binding energy per nucleon (CBE/A) for an object with a fixed energy (6.000 MeV) per bond. For this quick calculation, neither the type of bond nor the structure of these bonds comes into consideration. Simply stated, this is a

Nuclide	A	Z	N	EBE in MeV	EBE/A	# Of up quarks	# Of down quarks	# Of updown bonds, classical	Classical CBE/A
H2	2	1	1	2.225	1.11	3	3	3	9
He3	3	2	1	7.718	2.57	5	4	4	8
He4	4	2	2	28.296	7.07	6	6	6	9
He5	5	2	3	26.626	5.33	7	8	7	8.4
Li6	6	3	3	31.995	5.33	9	9	9	9
Li7	7	3	4	39.245	5.61	10	11	10	8.571
Be8	8	4	4	56.5	7.06	12	12	12	9
Be9	9	4	5	58.165	6.46	13	14	13	8.667
B10	10	5	5	64.751	6.48	15	15	15	9
B11	11	5	6	76.205	6.93	16	17	16	8.727
C12	12	6	6	92.162	7.68	18	18	18	9
C13	13	6	7	97.108	7.47	19	20	19	8.769
N14	14	7	7	104.659	7.48	21	21	21	9
N15	15	7	8	115.492	7.7	22	23	22	8.8
O16	16	8	8	127.619	7.98	24	24	24	9
O17	17	8	9	131.762	7.75	25	26	25	8.824
O18	18	8	10	139.808	7.77	26	28	26	8.667
F19	19	9	10	147.801	7.78	28	29	28	8.842
Ne20	20	10	10	160.65	8.03	30	30	30	9
Ne21	21	10	11	167.406	7.97	31	32	31	8.857
Ne22	22	10	12	177.77	8.08	32	34	32	8.727
Na23	23	11	12	186.564	8.11	34	35	34	8.87
Mg24	24	12	12	198.257	8.26	36	36	36	9
Mg25	25	12	13	205.587	8.22	37	38	37	8.88
Mg26	26	12	14	216.681	8.33	38	40	38	8.769
Al27	27	13	14	224.952	8.33	40	41	40	8.889
Si28	28	14	14	236.537	8.45	42	42	42	9
Si29	29	14	15	245.01	8.45	43	44	43	8.897
Si30	30	14	16	255.62	8.52	44	46	44	8.8
P31	31	15	16	262.917	8.48	46	47	46	8.903
S32	32	16	16	271.78	8.49	48	48	48	9
S33	33	16	17	280.422	8.5	49	50	49	8.909
S34	34	16	18	291.839	8.58	50	52	50	8.824
Cl35	35	17	18	298.21	8.52	52	53	52	8.914
S36	36	16	20	308.71	8.58	52	56	52	8.667
Ar36	36	18	18	306.716	8.52	54	54	54	9
Cl37	37	17	20	318.784	8.62	54	57	54	8.757
Ar38	38	18	20	327.343	8.61	56	58	56	8.842
K39	39	19	20	333.724	8.56	58	59	58	8.923
Ar40	40	18	22	343.81	8.6	58	62	58	8.7

Nuclide	A	Z	N	EBE in MeV	EBE/A	# Of up quarks	# Of down quarks	# Of updown bonds, classical	Classical CBE/A
Ca40	40	20	20	342.053	8.55	60	60	60	9
K41	41	19	22	351.619	8.58	60	63	60	8.78
Ca42	42	20	22	361.895	8.62	62	64	62	8.857
Ca43	43	20	23	369.828	8.6	63	66	63	8.791
Ca44	44	20	24	380.96	8.66	64	68	64	8.727
Sc45	45	21	24	387.849	8.62	66	69	66	8.8
Ca46	46	20	26	398.772	8.67	66	72	66	8.609
Ti46	46	22	24	398.194	8.66	68	70	68	8.87
Ti47	47	22	25	407.072	8.66	69	72	69	8.809
Ca48	48	20	28	415.992	8.67	68	76	68	8.5
Ti48	48	22	26	418.699	8.72	70	74	70	8.75
Ti49	49	22	27	426.841	8.71	71	76	71	8.694
Ti50	50	22	28	437.78	8.76	72	78	72	8.64
Cr50	50	24	26	435.047	8.7	74	76	74	8.88
V51	51	23	28	445.842	8.74	74	79	74	8.706
Cr52	52	24	28	456.345	8.78	76	80	76	8.769
Cr53	53	24	29	464.287	8.76	77	82	77	8.717
Cr54	54	24	30	474.009	8.78	78	84	78	8.667
Fe54	54	26	28	471.765	8.74	80	82	80	8.889
Mn55	55	25	30	482.075	8.77	80	85	80	8.727
Fe56	56	26	30	492.257	8.79	82	86	82	8.786
Fe57	57	26	31	499.905	8.77	83	88	83	8.737
Fe58	58	26	32	509.945	8.79	84	90	84	8.69
Ni58	58	28	30	506.456	8.73	86	88	86	8.897
Co59	59	27	32	517.314	8.77	86	91	86	8.746
Ni60	60	28	32	526.842	8.78	88	92	88	8.8
Zn70	70	30	40	611.08	8.73	100	110	100	8.571
Ge70	70	32	38	610.519	8.72	102	108	102	8.743
Se80	80	34	46	696.867	8.71	114	126	114	8.55
Kr80	80	36	44	695.438	8.69	116	124	116	8.7
Zr90	90	40	50	783.895	8.71	130	140	130	8.667
Ru100	100	44	56	861.929	8.62	144	156	144	8.64
Cd113	113	48	65	963.557	8.53	161	178	161	8.549
In113	113	49	64	963.091	8.52	162	177	162	8.602
Sn117	117	50	67	995.623	8.51	167	184	167	8.564
Xe129	129	54	75	1087.648	8.43	183	204	183	8.512
Ce142	142	58	84	1185.28	8.35	200	226	200	8.451
Nd142	142	60	82	1185.148	8.35	202	224	202	8.535
Sm150	150	62	88	1239.253	8.26	212	238	212	8.48
Gd150	150	64	86	1236.39	8.24	214	236	214	8.56

Nuclide	A	Z	N	EBE in MeV	EBE/A	# Of up quarks	# Of down quarks	# Of updown bonds, classical	Classical CBE/A
Dy162	162	66	96	1323.884	8.17	228	258	228	8.444
Yb172	172	70	102	1392.764	8.1	242	274	242	8.442
W183	183	74	109	1465.526	8.01	257	292	257	8.426
Pt194	194	78	116	1539.578	7.94	272	310	272	8.412
Au197	197	79	118	1559.397	7.92	276	315	276	8.406
Hg200	200	80	120	1581.207	7.91	280	320	280	8.4
Hg204	204	80	124	1608.65	7.89	284	328	284	8.353
Pb204	204	82	122	1605.343	7.87	286	326	286	8.412

Table 1.
A, Z, N, EBE, EBE/a, # up quarks, # down quarks, # possible bonds, and non-quantum calculated binding energy (classical CBE) for a representative sample of nuclides.

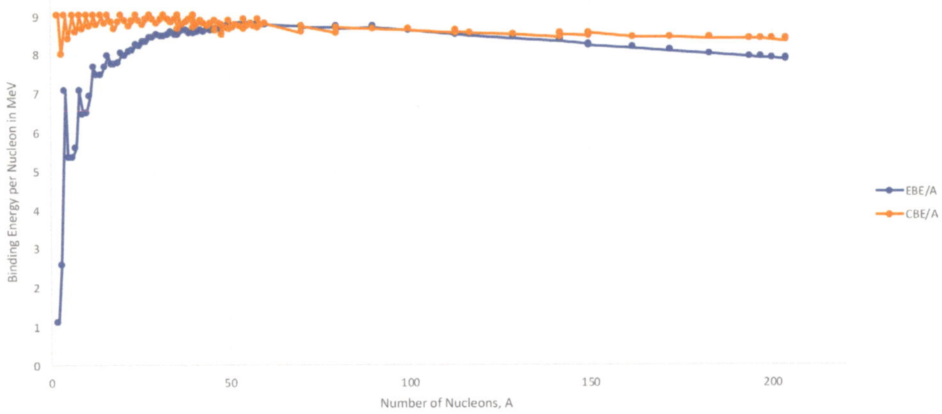

Figure 4.
In blue, a plot of the experimental binding energy (EBE) per nucleon. In orange, a plot of the calculated binding energy (CBE) per nucleon, based on the number of possible non-quantum up-to-down quark pairs and a fixed binding energy per bonded pair.

theory-independent calculation of the number of possible bonded pairs times a fixed binding energy per bonded pair.

4. Quantum considerations

A nucleus is a quantum object, and being so, certain quantum rules must apply. A known phenomenological feature of the nuclear force is the QCD hard-core repulsion. The hard-core repulsion states that nucleons, such as a proton or neutron, cannot overlap in their spatial location [11, 12]. If too many bonds are formed for either ^{2}H or ^{3}H or ^{3}He, overlap will occur. This overlap is illustrated in **Figure 5**.

To prevent this overlap, hydrogen ^{2}H can have only one bond instead of two or three. Similarly, helium ^{3}He (as well as hydrogen ^{3}H) can have only three bonds instead of four or five. Three other nuclides are subject to this constraint, those with odd-odd configurations: ^{6}Li, ^{10}B, and ^{14}N. Specifically, the odd neutron and the odd proton cannot bond twice to either each other or to another nucleon. Other stable nuclides are not affected by the application of this rule, since there are enough nucleons to prevent an overlap from occurring for the larger nuclides.

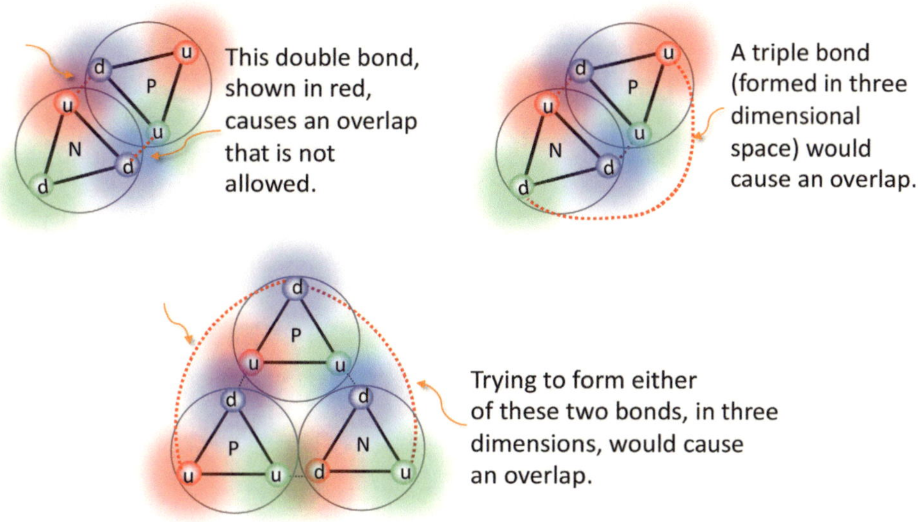

Figure 5.
An illustration of the overlapping nucleons if too many bonds are attempted. For the nuclide ^{2}H, if two or three bonds are attempted, indicated by the red bonds, an overlap occurs. Similarly, ^{3}He can form only three bonds. If four or five are attempted, an overlap occurs in three dimensions.

Quantum mechanics also states there can be no net electric dipole moment for the nuclide [13, 14]. For this second quantum rule, three more bonds must be subtracted from the number of bonds available, in order to remove the electric dipole moment. Without stating any specific configuration for the nuclide, this reduction of bonds can be best understood from the fact that the electric charge distribution of the nuclide must not have a net asymmetry in electrical charge for any of the three spatial dimensions, x, y, or z. To prevent an electric dipole moment, a bond is broken in each of these three dimensions, so that the net charge is symmetric about the x, y, and z axes. This quantum requirement removes three of the classically-allowed bonds. This rule applies to all stable nuclides, except for the three very smallest stable nuclides, ^{2}H, ^{3}He, and ^{4}He.

The inclusion of these two quantum rules is shown in **Table 2**. The first 8 columns of **Table 2** are similar to the first 8 columns of **Table 1**. Also shown in **Table 2** is the number of possible quantum bonds for each nuclide, taking into consideration the two above mentioned quantum rules. The last three columns of **Table 2** show the quantum calculated binding energy, the CBE/A, and the percent error of that calculated energy, as compared with the experimental binding energy.

As before for this simple calculation, the calculated binding energy is the number of bonds times a fixed energy per bond. The energy per bond is the only selected parameter; for this simple calculation, it is 6.000 MeV per bond. These plots take into consideration the quantum rules of hard-core repulsion and zero electric dipole moment. These data are plotted in **Figures 6** and 7. In **Figure 6**, a representative sample of all of the stable nuclides is shown, out to lead 204Pb. In **Figure 7**, only the first 60 nuclides are plotted, to show the detail. As before, when there is more than one stable nuclide for a given mass number, these additional points are plotted as well.

To reiterate, this is a very quick and easy calculation, only involving a simple numerical count of quantum-allowed internucleon up-to-down quark pairs. This calculation does not specify the arrangement of the nucleons or the mechanism of the bond. It is simply a count of the quantum-allowed up-to-down quark bonds.

Nuclide	A	Z	N	EBE in MeV	EBE/A	# Of up quarks	# Of down quarks	# Of updown bonds, classical	# Of updown bonds, quantum	Calculated binding energy (CBE) in MeV	CBE/A	%Error
H2	2	1	1	2.225	1.11	3	3	3	1	6	3	−169.66
He3	3	2	1	7.718	2.57	5	4	4	3	18	6	−133.22
He4	4	2	2	28.296	7.07	6	6	6	6	36	9	−27.23
He5	5	2	3	26.626	5.33	7	8	7	4	24	4.8	9.86
Li6	6	3	3	31.995	5.33	9	9	9	5	30	5	6.24
Li7	7	3	4	39.245	5.61	10	11	10	7	42	6	−7.02
Be8	8	4	4	56.5	7.06	12	12	12	9	54	6.75	4.42
Be9	9	4	5	58.165	6.46	13	14	13	10	60	6.667	−3.15
B10	10	5	5	64.751	6.48	15	15	15	11	66	6.6	−1.93
B11	11	5	6	76.205	6.93	16	17	16	13	78	7.091	−2.36
C12	12	6	6	92.162	7.68	18	18	18	15	90	7.5	2.35
C13	13	6	7	97.108	7.47	19	20	19	16	96	7.385	1.14
N14	14	7	7	104.659	7.48	21	21	21	17	102	7.286	2.54
N15	15	7	8	115.492	7.7	22	23	22	19	114	7.6	1.29
O16	16	8	8	127.619	7.98	24	24	24	21	126	7.875	1.27
O17	17	8	9	131.762	7.75	25	26	25	22	132	7.765	−0.18
O18	18	8	10	139.808	7.77	26	28	26	23	138	7.667	1.29
F19	19	9	10	147.801	7.78	28	29	28	25	150	7.895	−1.49
Ne20	20	10	10	160.65	8.03	30	30	30	27	162	8.1	−0.84
Ne21	21	10	11	167.406	7.97	31	32	31	28	168	8	−0.35
Ne22	22	10	12	177.77	8.08	32	34	32	29	174	7.909	2.12
Na23	23	11	12	186.564	8.11	34	35	34	31	186	8.087	0.3
Mg24	24	12	12	198.257	8.26	36	36	36	33	198	8.25	0.13

Nuclide	A	Z	N	EBE in MeV	EBE/A	# Of up quarks	# Of down quarks	# Of updown bonds, classical	# Of updown bonds, quantum	Calculated binding energy (CBE) in MeV	CBE/A	%Error
Mg25	25	12	13	205.587	8.22	37	38	37	34	204	8.16	0.77
Mg26	26	12	14	216.681	8.33	38	40	38	35	210	8.077	3.08
Al27	27	13	14	224.952	8.33	40	41	40	37	222	8.222	1.31
Si28	28	14	14	236.537	8.45	42	42	42	39	234	8.357	1.07
Si29	29	14	15	245.01	8.45	43	44	43	40	240	8.276	2.04
Si30	30	14	16	255.62	8.52	44	46	44	41	246	8.2	3.76
P31	31	15	16	262.917	8.48	46	47	46	43	258	8.323	1.87
S32	32	16	16	271.78	8.49	48	48	48	45	270	8.438	0.65
S33	33	16	17	280.422	8.5	49	50	49	46	276	8.364	1.58
S34	34	16	18	291.839	8.58	50	52	50	47	282	8.294	3.37
Cl35	35	17	18	298.21	8.52	52	53	52	49	294	8.4	1.41
S36	36	16	20	308.71	8.58	52	56	52	49	294	8.167	4.76
Ar36	36	18	18	306.716	8.52	54	54	54	51	306	8.5	0.23
Cl37	37	17	20	318.784	8.62	54	57	54	51	306	8.27	4.01
Ar38	38	18	20	327.343	8.61	56	58	56	53	318	8.368	2.85
K39	39	19	20	333.724	8.56	58	59	58	55	330	8.462	1.12
Ar40	40	18	22	343.81	8.6	58	62	58	55	330	8.25	4.02
Ca40	40	20	20	342.053	8.55	60	60	60	57	342	8.55	0.02
K41	41	19	22	351.619	8.58	60	63	60	57	342	8.341	2.74
Ca42	42	20	22	361.895	8.62	62	64	62	59	354	8.429	2.18
Ca43	43	20	23	369.828	8.6	63	66	63	60	360	8.372	2.66
Ca44	44	20	24	380.96	8.66	64	68	64	61	366	8.318	3.93
Sc45	45	21	24	387.849	8.62	66	69	66	63	378	8.4	2.54

Nuclide	A	Z	N	EBE in MeV	EBE/A	# Of up quarks	# Of down quarks	# Of updown bonds, classical	# Of updown bonds, quantum	Calculated binding energy (CBE) in MeV	CBE/A	%Error
Ca46	46	20	26	398.772	8.67	66	72	66	63	378	8.217	5.21
Ti46	46	22	24	398.194	8.66	68	70	68	65	390	8.478	2.06
Ti47	47	22	25	407.072	8.66	69	72	69	66	396	8.426	2.72
Ca48	48	20	28	415.992	8.67	68	76	68	65	390	8.125	6.25
Ti48	48	22	26	418.699	8.72	70	74	70	67	402	8.375	3.99
Ti49	49	22	27	426.841	8.71	71	76	71	68	408	8.327	4.41
Ti50	50	22	28	437.78	8.76	72	78	72	69	414	8.28	5.43
Cr50	50	24	26	435.047	8.7	74	76	74	71	426	8.52	2.08
V51	51	23	28	445.842	8.74	74	79	74	71	426	8.353	4.45
Cr52	52	24	28	456.345	8.78	76	80	76	73	438	8.423	4.02
Cr53	53	24	29	464.287	8.76	77	82	77	74	444	8.377	4.37
Cr54	54	24	30	474.009	8.78	78	84	78	75	450	8.333	5.07
Fe54	54	26	28	471.765	8.74	80	82	80	77	462	8.556	2.07
Mn55	55	25	30	482.075	8.77	80	85	80	77	462	8.4	4.16
Fe56	56	26	30	492.257	8.79	82	86	82	79	474	8.464	3.71
Fe57	57	26	31	499.905	8.77	83	88	83	80	480	8.421	3.98
Fe58	58	26	32	509.945	8.79	84	90	84	81	486	8.379	4.7
Ni58	58	28	30	506.456	8.73	86	88	86	83	498	8.586	1.67
Co59	59	27	32	517.314	8.77	86	91	86	83	498	8.441	3.73
Ni60	60	28	32	526.842	8.78	88	92	88	85	510	8.5	3.2
Zn70	70	30	40	611.08	8.73	100	110	100	97	582	8.314	4.76
Ge70	70	32	38	610.519	8.72	102	108	102	99	594	8.486	2.71
Se80	80	34	46	696.867	8.71	114	126	114	111	666	8.325	4.43

DOI: http://dx.doi.org/10.5772/intechopen.94377

Nuclide	A	Z	N	EBE in MeV	EBE/A	# Of up quarks	# Of down quarks	# Of updown bonds, classical	# Of updown bonds, quantum	Calculated binding energy (CBE) in MeV	CBE/A	%Error
Kr80	80	36	44	695.438	8.69	116	124	116	113	678	8.475	2.51
Zr90	90	40	50	783.895	8.71	130	140	130	127	762	8.467	2.79
Ru100	100	44	56	861.929	8.62	144	156	144	141	846	8.46	1.85
Cd113	113	48	65	963.557	8.53	161	178	161	158	948	8.389	1.61
In113	113	49	64	963.091	8.52	162	177	162	159	954	8.442	0.94
Sn117	117	50	67	995.623	8.51	167	184	167	164	984	8.41	1.17
Xe129	129	54	75	1087.648	8.43	183	204	183	180	1080	8.372	0.7
Ce142	142	58	84	1185.28	8.35	200	226	200	197	1182	8.324	0.28
Nd142	142	60	82	1185.148	8.35	202	224	202	199	1194	8.408	−0.75
Sm150	150	62	88	1239.253	8.26	212	238	212	209	1254	8.36	−1.19
Gd150	150	64	86	1236.39	8.24	214	236	214	211	1266	8.44	−2.39
Dy162	162	66	96	1323.884	8.17	228	258	228	225	1350	8.333	−1.97
Yb172	172	70	102	1392.764	8.1	242	274	242	239	1434	8.337	−2.96
W183	183	74	109	1465.526	8.01	257	292	257	254	1524	8.328	−3.99
Pt194	194	78	116	1539.578	7.94	272	310	272	269	1614	8.32	−4.83
Au197	197	79	118	1559.397	7.92	276	315	276	273	1638	8.315	−5.04
Hg200	200	80	120	1581.207	7.91	280	320	280	277	1662	8.31	−5.11
Hg204	204	80	124	1608.65	7.89	284	328	284	281	1686	8.265	−4.81
Pb204	204	82	122	1605.343	7.87	286	326	286	283	1698	8.324	−5.77

Table 2.
A representative sample of nuclides, showing quantum-allowed bonded up-to-down quark pairs for each nuclide.

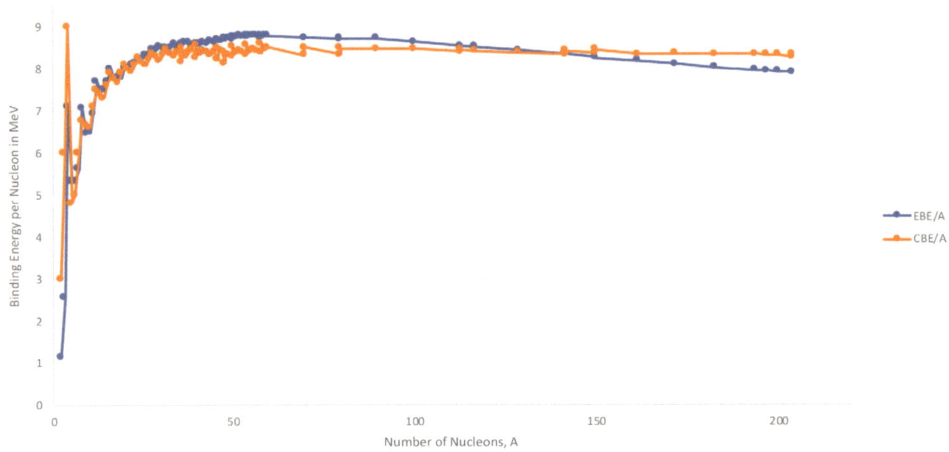

Figure 6.
A plot of the experimental nuclear binding energy per nucleon (blue) and the simple quantum calculated binding energy (orange).

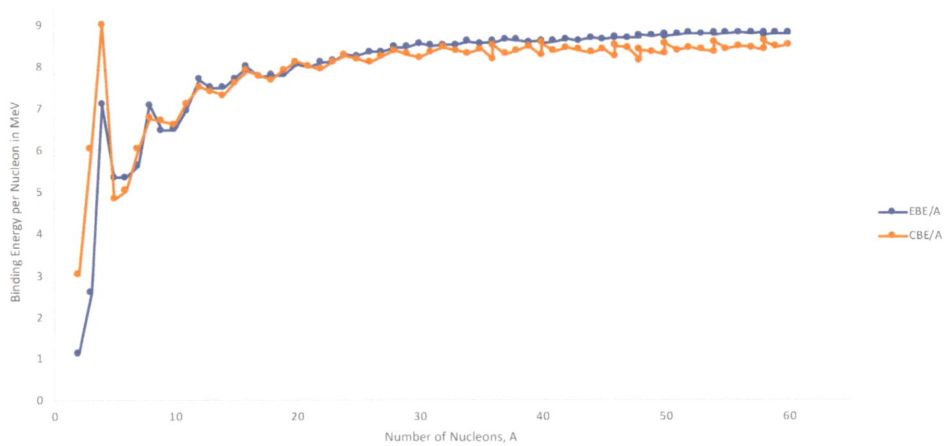

Figure 7.
A plot of the experimental nuclear binding energy per nucleon (blue) and the simple quantum calculated binding energy (orange) for only the first 60 nuclides.

Other than being quantum, this calculation is theory independent, and as such, it is not subject to theoretical criticisms or theoretical differences of opinion.

5. Discussion

The excellent reproduction of the experimental data for these calculated results is impressive, especially considering that there is only one empirically-selected variable for this calculation, the value of 6.000 MeV for the bond energy, instead of the five empirically-selected variables for the Weizsäcker formula. This reproduction of the experimental data is especially impressive considering that other currently accepted nuclear theories cannot easily duplicate this curve.

In terms of the possible mechanism for the bond, the residual chromodynamic force between the color charges for internucleon quark-to-quark bonding is one possibility. Another possibility for this bond becomes apparent when it is recalled that the up quark has an electric charge +2/3 the charge of a proton, the down quark has an electric charge of −1/3 the charge of a proton, and both quarks carry a magnetic moment. These electromagnetic properties of the up and down quarks create a strong attractive electromagnetic force between the up and the down quarks; the strength of this electromagnetic force is dependent only on the minimum proximity between the up and down quarks engaged in the bond. Historically, it was believed that the strength of the electromagnetic force had an upper limit, based on the misconceptions that protons were homogeneously charged and that quarks did not exist. However, these misconceived notions are invalid when quarks, which contain all of the electric charge for the nucleons, are taken into consideration.

The internuclear quark-to-quark bond is most likely some combination of both the electromagnetic charge and the color charge of the quarks, but the relative percentages of these two contributions is not postulated here. Regardless of the relative percentages, the electromagnetic component of this bond should not be ignored–as is usually the case in current theories. When any internucleon quark-to-quark bond is considered, the electromagnetic component must be taken into full account, rather than being considered relatively unimportant. A more detailed analysis of the electromagnetic contribution of this internucleon up-to-down quark bonding can easily be made by using the standard electromagnetic Eqs. A detailed analysis would include the addition of the energy due to all electric charges interacting with each other. In other words, this would be a double summation of the interaction for each electric charge of each quark with every other electric charge on all other quarks [15]. This double summation calculation would inherently include the Coulomb energy of the net repulsive electric energies among the protons.

Similarly, a more detailed electromagnetic analysis would also include the variation of the electromagnetic bond due to the vector orientation of the magnetic moments of the quarks. The energy of the magnetic moments interacting with each other should be included, which again would be a double summation for the magnetic interaction for all of the magnetic moment vectors [16]. Finally, the kinetic energy of the quantum spin of the nuclide should also be included in this more detailed binding energy calculation [17, 18]. However, for this more detailed and accurate calculation to be done, the lowest energy configuration of the nuclide must be determined and specified before the electromagnetic interaction energies can be accurately calculated.

6. Conclusion

An extremely simple calculation of the internucleon up-to-down quark bonding has been made, giving excellent results in duplicating the nuclear binding energy curve, using only one parameter rather than five. The resulting errors for nuclides going up to lead ^{204}Pb are only few percent. The average error, going from A = 10 to A = 60, is only 2.32% with a standard deviation for that error of only 1.91%. Also, due to the inherent similarities of this concept to the currently accepted residual chromodynamic force model, with its quark-to-quark internucleon bonding, the existence of an internucleon up-to-down quark bond cannot be relegated as implausible.

An obvious implication of these results is that a significant part of the nuclear force is electromagnetic. To some, this may be an unexpected implication, but not unfeasible, especially when the electromagnetic attraction of the up-to-down quarks is considered. If one only considers, as is the case historically, that homogenously charged protons cannot bond to other homogeneously charged protons, then the concept that the nuclear force could be partially electromagnetic is deemed implausible. However, with the understanding that the electrical charges of the up and down quarks are able to attract each other and bond to each other, and given that the RCDF allows a quark-to-quark internucleon bond to occur, such restrictions about the nuclear force being partly electromagnetic are no longer relevant.

The excellent reproduction of experimental binding energy data with only one empirically-selected variable strongly suggests that the internucleon up-to-down quark bonding is a concept that should be seriously considered and more thoroughly examined by nuclear physicists.

Author details

Nancy Lynn Bowen
Colorado Mountain College, Glenwood Springs, CO, USA

*Address all correspondence to: nbowen@coloradomtn.edu

References

[1] Gamow G. Gamow's description of the liquid drop model. Royal Society. 1929;123:373-390.

[2] VonWeizsäcker C. Zur theorie der kernmassen. *Zeitschrift für Physik*. 1935; 96:431-458.

[3] Basdevant J, Rich J, Spiro M. Fundamentals in Nuclear Physics. New York: Springer Science + Business Media, LLC; 2005. 67 p.

[4] Greenberg O. Color Charge Degree of Freedom in Particle Physics. In: Greenberger D, Hentschel K, Weinert F, editors. Compendium of Quantum Physics. Berlin, Heidelberg: Springer; 2009. p. 109-111. DOI: https://doi.org/10.1007/978-3-540-70626-7_32

[5] Eisberg R, Resnick R. Quantum Physics of Atoms, Molecules, Solids, Nuclei, and Particles. New York: Wiley; 1985. 683 p.

[6] Meisner U. Modern theory of nuclear forces AD 2006. European Physics Journal. 2007;A31:397-402. DOI: 10.1140/epja/i2006-10170-1.

[7] Machleidt R. Nuclear Forces. Scholarpedia, 9(1):30710, 2014. DOI: 10: 4249/scholarpedia.30710.

[8] Duane S, Kennedy A, Pendleton B, Roweth D. Physics Letters B. 1987;195: 216.

[9] Griffiths J. Introduction to Elementary Particle. New York: John Wiley & Sons; 1987. 47 p.

[10] National Nuclear Data Center. NuDat 2 database [Internet]. 2020. Available from: http://www.nndc.bnl.gov/nudat2/ [Accessed 2019]

[11] Wilczek F. Particle physics: hard-core revelations. Nature. 2007;445: 156-157. DOI: 10.1038/445156a

[12] Ishii N, Aoki S, Hatsuda T. Nuclear force from lattice QCD. Physics Review Letters. 2007;99:022001. DOI: 10.1103/PhysRevLett.99.022001

[13] McGervey J. Introduction to Modern Physics. New York: Academic Press; 1971. 479 p.

[14] Bertulani C. Nuclear Physics in a Nutshell. Princeton, New Jersy: Princeton University Press; 2007. 105 p.

[15] Owen G. Electromagnetic Theory. Boston: Allen and Bacon; 1963. 23 p.

[16] Yosida K. Theory of Magnetism. New York: Springer; 1996. 13 p.

[17] Krane K. Introductory Nuclear Physics. New York: Wiley; 1988. 144 p.

[18] McGervey J. Introduction to Modern Physics. New York: Academic Press; 1971. 498 p.

Chapter 4

Spectrosopic Study of Baryons

Zalak Shah and Ajay Kumar Rai

Abstract

Baryons are the combination of three quarks(antiquarks) configured by qqq($\overline{qqq}$). They are fermions and obey the Pauli's principal so that the total wave function must be anti-symmetric. The SU(5) flavor group includes all types of baryons containing zero, one, two or three heavy quarks. The Particle Data Froup (PDG) listed the ground states of most of these baryons and many excited states in their summary Table. The radial and orbital excited states of the baryons are important to calculate, from that the Regge trajectories will be constructed. The quantum numbers will be determined from these slopes and intersects. Thus, we can help experiments to determine the masses of unknown states. The other hadronic properties like decays, magnetic moments can also play a very important role to emphasize the baryons. It is also interesting to determine the properties of exotic baryons nowadays.

Keywords: baryons, potential model, mass spectra, Regge trajectories, decays

1. Introduction

The particle physics has a remarkable track record of success by discovering the basic building blocks of matter and their interactions. Everything in the observed universe is found to be made from a few basic building blocks called *fundamental particles*, governed by four fundamental forces. All of these are encapsulated in the **Standard Model (SM)**. The Standard Model has been established in early 1970s and so far it is the most precise theory ever made by mankind. In the Standard Model elementary particles are considered to be the constituents of all observed matter. These elementary particles are quarks and leptons and the force carrying particles, such as gluons and W, Z bosons. All stable matter in the universe is made from particles that belong to the first generation; any heavier particles quickly decay to the next most stable level. The concept of the quark was first proposed by Murray Gell-Mann and George Zweig in 1937. The quarks have very strong interaction with each other, that is a reason, they always stuck inside composite system. But quarks interact with leptons weakly, the best example of this is protons, neutrons and electrons of atomic nuclei. The flavored quarks combine together in various aggregates called hadrons. The Standard Model includes 12 elementary particles of half-spin known as fermions. They follow the Pauli exclusion principle and each of them have a corresponding anti-particle. There are also twelve integer-spin particles which mediate interactions between these particles known as Bosons. They obey the Bose-Einstein statistics. The classification of bosons and fermions along with hadrons are drawn in **Figure 1**.

The Quantum Chromodynamics (QCD) as the theory of strong interactions was successfully used to explore spectroscopic parameters and decay channels of hadrons during last five decades. The interaction is governed by massless spin 1 objects

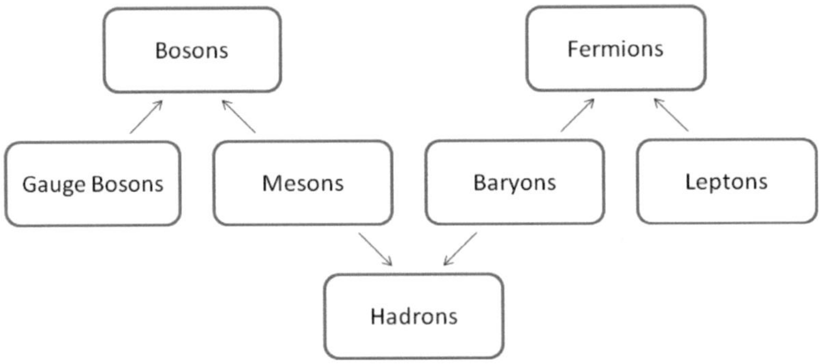

Figure 1.
The classifications of particles.

called gluons. Quarks inside the hadrons exchange gluons and create a very strong color force field. To conserve color charge, quarks constantly change their color by exchanging gluons with other quarks. As the quarks within a hadron get closer together, the force of containment gets weaker so that it asymptotically approaches zero for close confinement. The quarks in close confinement are completely free to move about. This condition is referred to as "asymptotic freedom". An essential requirements for the progress in hadronic physics is the full usage of present facilities and development of new ones, with a clear focous on experiments that provide geniune insight into the inner workings of QCD.

Hadron spectroscopy is a tool to reveal the dynamics of the quark interactions within the composite systems. The short-lived hadrons and missing excited states could be identified through the possible decays of the resonance state. The experimentally discovered states are listed in summary tables of Particle Data Group [1]. The worldwide experiments such as LHCb, BELLE, BARBAR, CDF, CLEO are main source of identification of heavy baryons so far and especially LHCb and Belle experiments have provided the new excited states in heavy baryon sector recently [2]. The various phenomenological approaches for spectroscopy is all about to use the potential and establish the excited resonances. These approches are, relativistic quark model, HQET, QCD Sum rules, Lattice-QCD, Regge Phenomology, and many Phenomenological models [3–8]. An overview to the current status of research in the field of baryon physics from an experimental and theoretical aspects with a view to provide motivation and scope for the present chapter. The present study covers the baryons with one heavy and two light quarks [9–13]; two heavy and one light quark [14, 15] as well as three heavy quarks [16–18]. We also like to discuss the spectroscopy of nucleons [19, 20]. The decay properties, magnetic moments and Regge Trajectories are also discussed.

2. Light, heavy flavored baryons and exotics

In the case of baryons, when three same quark combines, definitely their electric charge, spin, orbital momentum would be same. This might violates the Pauli exclusion principal stated *"no two identical fermions can be found in the same quantum state at the same time"*. However, the color quantum number of each quark is different so that the exclusion principle would not be violated. One of the most significant aspects of the baryon spectrum is the existence of almost degenerate levels of different charges which have all the characteristics of isospin multiplets, quartets, triplets, doublets and singlets. A more general charge formula that encompasses all these nearly degenerate multiplets is

$$Q = I_3 + \frac{1}{2}Y, \qquad Y = B + S + C/B' \tag{1}$$

where Q, I_3, Y, B, C, S, B′ are referred as charge, isospin, hypercharge, baryon number, charm, strangeness and bottomness,respectively. The strangeness of baryon is always negative.

The baryons are strongly interacting fermions made up of three quarks and have $\frac{1}{2}$ integer spin. They obey the Pauli exclusion principle, thus the total wave function must be anti-symmetric under the interchange of any two quarks. Since all observed hadrons are color singlets, the color component of the wave function must be completely anti-symmetric.

$$\text{For Octet,} \quad \begin{cases} \vec{s}_u \cdot \vec{s}_s + \vec{s}_d \cdot \vec{s}_s = -1 & \text{symmetric} \\ \vec{s}_u \cdot \vec{s}_s + \vec{s}_d \cdot \vec{s}_s = 0 & \text{antisymmetric} \end{cases}$$

$$\text{For Decuplet,} \quad \begin{cases} \vec{s}_u \cdot \vec{s}_s + \vec{s}_d \cdot \vec{s}_s = \frac{1}{2} & \text{symmetric} \\ \vec{s}_u \cdot \vec{s}_s + \vec{s}_d \cdot \vec{s}_s = \frac{3}{2} & \text{antisymmetric} \end{cases}$$

Here, $\vec{s}_u$, $\vec{s}_d$ and $\vec{s}_s$ are spin of u, d, s quarks.

It was considered that u, d and s are the sole elementary quarks. The symmetry group to consider three flavors of quark is done by SU(3) symmetry group. SU(3) flavor symmetry of light quarks. Each of these symmetries refers to an underlying threefold symmetry in strong interaction physics. SU(3) is the group symmetry transformations of the 3-vector wavefunction that maintain the physical constraint that the total probability for finding the particle in one of the three possible states equals 1. A Young diagram is the best way to represnts the symmetriesconsists of an array of boxes arranged in one or more left-justified rows, with each row being at least as long as the row beneath. The correspondence between a diagram and a multiplet label is: The top row juts out α boxes to the right past the end of the second row, the second row juts out β boxes to the right past the end of the third row, etc. The representation is shown in **Figures 2** and **3**.

The flavor wave functions of baryon states can then be constructed to be members of SU(3) multiplets as [5].

$$3 \otimes 3 \otimes 3 = 10_S \oplus 8_M \oplus 8_M \oplus 1_A$$

Murray Gell-Mann introduced the Eightfold geometrical pattern for mesons and baryons in 1962 [21]. The eight lighest baryons fit into the hexagonal array with two particle in center are called baryon octet. A triangular array with 10 particles are called the baryon decuplet. Moreover, the antibaryon octet and decuplet also exist

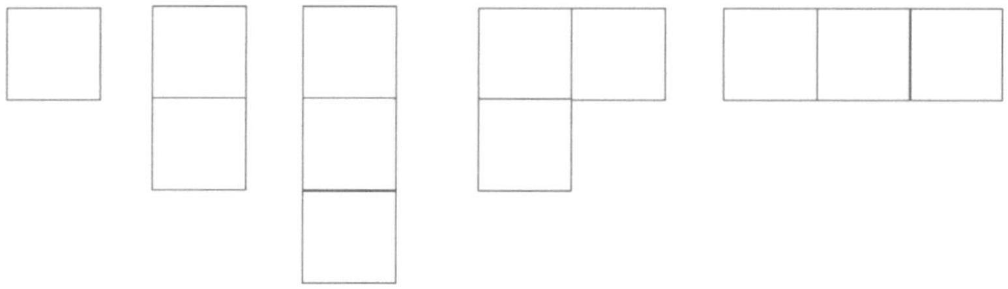

Figure 2.
The representation of the multiplets (1,0), (0,1), (0,0), (1,1), and (3,0) in SU(3) diagrams.

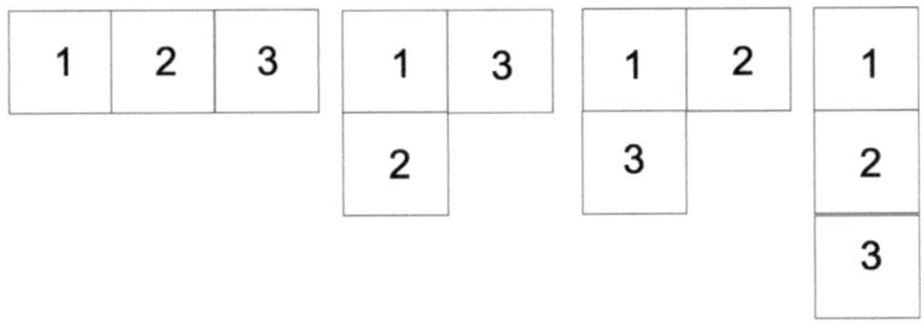

Figure 3.
A standard young tableaux for N = 3.

with opposite charge and strangeness [22]. Baryons having only u and d quarks are called nucleons N and Δ resonances. The *proton* and *neutron* have spin ($I_3 = \frac{1}{2}$) and Δ particles have spin ($I_3 = \frac{3}{2}$). The four possible four combinations of the symmetric wave function gives four Δ particles; Δ^{++} (uuu, $I_3 = 3/2$), Δ^{+} (uud, I_3 =1/2), Δ^{0} (udd, I_3 =1/2) and Δ^{-} (ddd, I_3 =3/2). Particles with combination of u, d and s quarks are called hyperons; Λ, Σ, Ξ and Ω. While discussing heavy sector baryons, we need baryons having heavy quark(s) combination. Any s quark(s) of hyperon baryons can be replaced by heavy quark (c, b) in heavy baryon particles. The added heavy quark(s) will be added to the suffix of the particular baryon. The particles are also depend on isospin quantum number, such that Σ and Ξ baryons have isospin triplets and doublets, respectively. Λ_c and Σ_c (Λ_b and Σ_b) are formed by replacing one s quark. For Ξ baryon, replacement of one s quark gives Ξ_c (Ξ_b) and the particles Ξ_{cc}, Ξ_{bb} and Ξ_{bc} found while replacing two s quarks. The biggest family is found for Ω particle. Replacement of one s quark gives Ω_c (Ω_b); two s quarks replace to provide Ω_{cc}, Ω_{bb} and Ω_{bc}; all three quark replacement with s quark give Ω_{ccc}, Ω_{bbb}, Ω_{bbc} and Ω_{ccb} particles.

SU(4) group includes all of the baryons containing zero, one, two or three heavy Q (charm or beauty) quarks with light u, d and s quarks. The number of particles in a multiplet, N=N(α,β,γ) is

$$N = \frac{(\alpha+1)}{1}\cdot\frac{(\beta+1)}{1}\cdot\frac{(\gamma+1)}{1}\cdot\frac{(\alpha+\beta+2)}{2}\cdot\frac{(\beta+\gamma+2)}{2}\cdot\frac{(\alpha+\beta+\gamma+3)}{3} \tag{2}$$

It is clear from Eq. (2) that multiplets that are conjugate to one another have the same number of particles, but so can other multiplets. The multiplets (3,0,0) and (1,1,0) each have 20 particles. This multiplet structure is expected to be repeated for every combination of spin and parity which provides a very rich spectrum of baryonic states. The multiplet numerology of the tensor product of three fundamental representation is given as:

$$4 \otimes 4 \otimes 4 = 20 \oplus 20'_1 \oplus 20'_2 \oplus \overline{4}. \tag{3}$$

Representation shows totally symmetric 20-plet, the mixed symmetric 20′-plet and the total anti-symmetric $\overline{4}$ multiplet. The charm baryon multiplets are presented in **Figure 4**. The ground levels of SU(4) group multiplets are SU(3) decuplet, octet, and singlet, respectively. These baryon states can be further decomposed according to the heavy quark content inside. According to the symmetry, the heavy baryons belong to two different SU(3) flavor representations: the symmetric sextet 6_s and anti-symmetric anti-triplet $\overline{3}_A$. It can also be represented by young tableaux (refer **Figure 2**). The observed resonances of all light and heavy baryons are listed in PDG (2020) baryon summary **Table 1**.

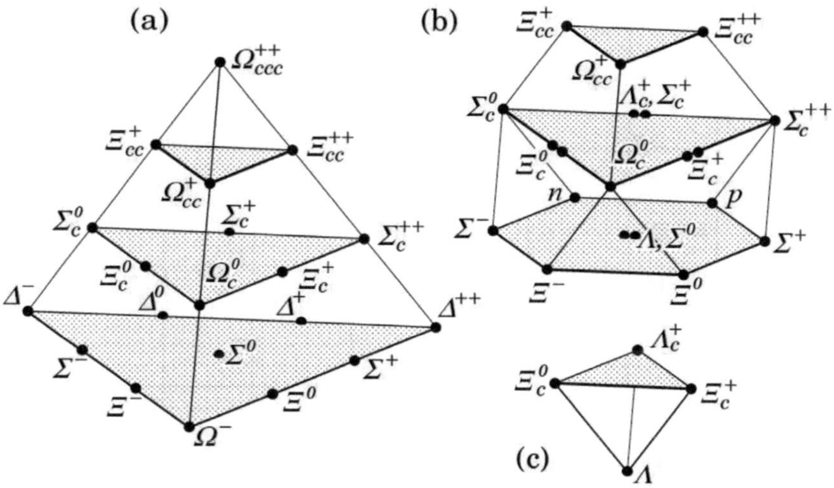

Figure 4.
The baryon multiplets in SU(4) group. (a) the spin 3/2 20-plet, (b) the spin 1/2 20′-plet and (c) $\bar{4}$ singlet [1].

Names	Exp. Mass (GeV) [1]	Predicted Mass (GeV)	Baryon State
$\Lambda_c^+(2765)218083$	2766.6±2.4	2678	$(2^2S_{\frac{1}{2}})$
$\Lambda_c^+(2940)$	$2939.3^{+1.4}_{-1.5}$	2927,2910	$(2^2P_{\frac{1}{2}}),(2^2P_{\frac{3}{2}})$
$\Sigma_c^{++}(2800)$	2801^{+4}_{-6}	2856,2791	$(1^2P_{\frac{3}{2}}),(1^4P_{\frac{5}{2}})$
$\Sigma_c^+(2800)$	2792^{+14}_{-5}	2805,2783	$(1^2P_{\frac{3}{2}}),(1^4P_{\frac{5}{2}})$
$\Sigma_c^0(2800)$	$2806 \pm 8 \pm 10$	2799,2778	$(1^2P_{\frac{3}{2}}),(1^4P_{\frac{5}{2}})$
$\Xi_c(2930)$	2931±3±5	2750,2733	$(1^2P_{\frac{3}{2}}),(1^2P_{\frac{5}{2}})$
$\Xi_c^+(2970)$	2971.4±3.3	2944,3059	$(2^2S_{\frac{1}{2}}),(2^4S_{\frac{3}{2}})$
$\Xi_c^0(2970)$	2968.0±2.6	2903,2859	$(2^2S_{\frac{1}{2}}),(2^4S_{\frac{3}{2}})$
$\Xi_c^+(3055)$	3054.2±1.2±0.5	3088	$(1^2D_{\frac{3}{2}})$
$\Xi_c^+(3080)$	3077.0±0.4	3062	$(1^2D_{\frac{5}{2}})$
$\Xi_c^0(3080)$	3079.9±1.4	3162,2997	$(1^2D_{\frac{3}{2}}),(1^4D_{\frac{3}{2}})$
$\Xi_c^+(3123)$	3122.9±1.3±0.3	3039	$(1^2D_{\frac{7}{2}})$
$\Xi_{cc}^+(3520)$	3518.9±0.9	3520	$(1^2S_{\frac{1}{2}})$

Table 1.
Our predicted baryonic states are compared with experimental unknown (J^P) excited states.

2.1 Exotics: non-conventional hadrons

Apart from the simplest pairings of quarks-anti quarks in formation of mesons and baryons, there are many observed states that do not fit into this picture. Numerous states have recently been found and some of those have exotic quark structures. Some of these exotics states are experimentally explained as *tetraquarks* (contains two-quarks and two anti-quarks) and *pentaquarks* (contains four-quarks plus an anti-quark) states with active gluonic degrees of freedom (hybrids), and even states of pure glue (glueballs) so far. Many experiments Belle, Barabar, CLEO, BESIII, LHCb, ATLAS, CMS and DO collaborations are working on the investigation of these exotic states. The theoretical approaches such as effective field theories of QCD, various quark moels, Sum rules, Lattice QCD, etc. also preicted many states of the exotic states.

The two pentaquarks $P_c^+(4380)$ and $P_c^+(4450)$, discovered in 2015 by the LHCb collaboration, in the $J/\psi pK$ invarnt mass distribution [23]. The newly observed states, $P_c^+(4440)$, $P_c^+(4457)$, $P_c^+(4312)$ were investigated via different methods in 2019 by LHCb. These states are considered in various recent studies and the majority suggested as negative spin parity quantum number [24]. The investigations of pentaquark states resulted in support of different possibilities for their sub-substructures leaving their structures still ambiguous. Therefore to discriminate their sub-structure we need further theoretical and experimental investigations.

R. Jaffe obtained six-quark states built of only light u, d, and s quarks called as *dibaryon* or *hexaquark* that belong to flavor group $SU_f(3)$. Using for analysis the MIT quark-bag model, Jaffe predicted existence of a H-dibaryon, i.e., a flavor-singlet and neutral six-quark uuddss bound state with isospin–spin-parity $I(J^P) = 0(0^+)$ [25]. In the past fifteen years, new states have been observed called the XYZ states, different from the ordinary hadrons. Some of them, like the charged states, are undoubtedly exotic. Theoretical study include the phenomenological quark model to exotics, non-relativistic effective field theories and lattice QCD calculations and enormous experimental studies we can see on XYZ states.

As a hadronic molecule [26, 27], deuteron has been well-established loosely bound state of a proton and a neutron. Ideally, the large masses of the heavy baryons reduce the kinetic energy of the systems, which makes it easier to form bound states. Such a system is approximately non- relativistic. Therefore, it is very interesting to study the binding of two heavy baryons *dibaryon* and a combination of heavy baryon and an antibaryon *baryonium*.

3. Specroscopic properties

Hadron spectroscopy is a key to strong interactions in the region of quark confinement and very useful for understanding the hadron as a bound state of quarks and gluons. Any system within a standard model becomes difficult to deal considering all the interaction of quark-quark, quark-gluon and gluon-gluon. This is the reason for using constituent quark mass incorporating all the other effects in the form of some parameters. The bound state heavy baryons can be studied in the QCD motivated potential models treating to the non relativistic Quantum mechanics. A Constituent Quark Model is a modelization of a baryon as a system of three quarks or anti-quarks bound by some kind of confining interaction. The present study deals with the Hypercentral Constituent Quark Model(hCQM), an effective way to study three body systems is through consideration of Jacobi coordinates.

The hypercentral approach has been applied to solve bound states and scattering problems in many different fields of physics and chemistry. The basic idea of the hypercentral approach to three-body systems is very simple. The two relative coordinates are rewritten into a single six dimensional vector and the non-relativistic *Schrödinger* equation in the six dimensional space is solved. The potential expressed in terms of the hypercentral radial co-ordinate, takes care of the three body interactions effectively [28]. The ground states and some of the excited states of light baryons has also been affirmed theoretically by this scheme. It is interesting to identify the mass spectrum of heavy baryons(singly, doubly and triply) in charm as well as bottom sector and then to the light sector. We consider a nonrelativistic Hamiltonian given by

$$H = \frac{P_x^2}{2m} + V(x) \tag{4}$$

where, $m = \frac{2m_\rho m_\lambda}{m_\rho + m_\lambda}$, is the reduced mass and m_ρ and m_λ are reduced masses of Jacobi co-ordinates $\vec{\rho}$ and $\vec{\lambda}$. x is the six dimensional radial hyper central coordinate of the three body system. Non-relativistically, this interaction potential, $V(x)$ consists of a central term $V_c(r)$ and spin dependent part $V_{SD}(r)$. The central part $V_c(r)$ is given in terms of vector (Couloumb) plus scalar (confining) terms as

$$V_c(r) = V_V + V_S = -\frac{2}{3}\frac{\alpha_s}{r} + \beta r^\nu \tag{5}$$

The short-distance part of the static three-quark system, arising from one-gluon exchange within baryon, is of Coulombic shape. Here, we can observe that the strong running coupling constant (α_s) becomes smaller as we decrease the distance, the effective potential approaches the lowest order one-gluon exchange potential given in Eq. (4) as r $\rightarrow$ 0. So, for short distances, one can use the one gluon exchange potential, taking into account the running coupling constant α_s. We employ the Coulomb plus power potential (CPP$_\nu$) with varying power index ν, as there is no definite indication for the choices of ν that works at different hadronic sector. The values of potential index ν is varying from 0.5 to 2.0; in other words, S.R (1/2), linear (1.0), 3/2 power- law (1.5) and quadratic (2.0) potentials are taken into account in case of singly heavy baryon. The hypercentral potential V(x) as the color coulomb plus power potential with first order correction and spin-dependent interactionare written as,

$$V(x) = V^0(x) + \left(\frac{1}{m_\rho} + \frac{1}{m_\lambda}\right) V^{(1)}(x) + V_{SD}(x) \tag{6}$$

where $V^0(x)$ is given by the sum of hyper Coulomb(hC) interaction and a confinement term

$$V^{(0)}(x) = \frac{\tau}{x} + \beta x^\nu \tag{7}$$

and first order correction as similar to the one given by [29],

$$V^{(1)}(x) = -C_F C_A \frac{\alpha_s^2}{4x^2} \tag{8}$$

Here, C_F and C_A are the Casimir charges of the fundamental and adjoint representation. For computing the mass difference between different degenerate baryonic states, we consider the spin dependent part of the usual one gluon exchange potential (OGEP). The spin-dependent part, $V_{SD}(x)$ contains three types of the interaction terms, such as the spin–spin term $V_{SS}(x)$, the spin-orbit term $V_{\gamma S}(x)$ and tensor term $V_T(x)$. Considering all isospin splittings, the ground and excited state masses are determined for all heavy baryon system. The radial excited states from 2S–5S and orbital excited states from 1P-5P, 1D-4D and 1F–2F are calculated using the hCQM formalism. These mass spectra can be found in our Refs. [10–18].

The correction used in the potential have its dominant effect on potential energy. In our calculation, the effect of the correction to the potential energy part is decreasing as mass of the system increasing. For example, if noticed the maximum effect of heavy Ξ baryon family then, for the radial excited states of Ξ baryons are $\Xi_c(3.0\%) > \Xi_{cc}(3.0\%) > \Xi_b(0.41\%) > \Xi_{bc}(0.3\%) > \Xi_{bb}(0.16\%)$. The orbital excited states are (1.4–3.5%) for Ξ_c, (0.1–0.4%) for Ξ_b and in case of doubly heavy region the error is rising in order of baryons $\Xi_{bb} > \Xi_{bc} > \Xi_{cc}$. Singly heavy baryons show the effect from (0.1–1.7%), Doubly heavy baryons show the effect from (0.1–1.0%) and Triply heavy baryons show the effect from (0.2–0.9%) in orbital excited states. For better idea, we shown the effect of masses with and without first order corrections in case of Σ_c baryon (See, **Figure 5**). In the Figure, we plotted the graph of potenial index vs. mass. The radial excited states 2S–5S and the orbital excited states 1P-5P are shown for Σ_c^{++}, Σ_c^0 and Σ_c^+ baryons.

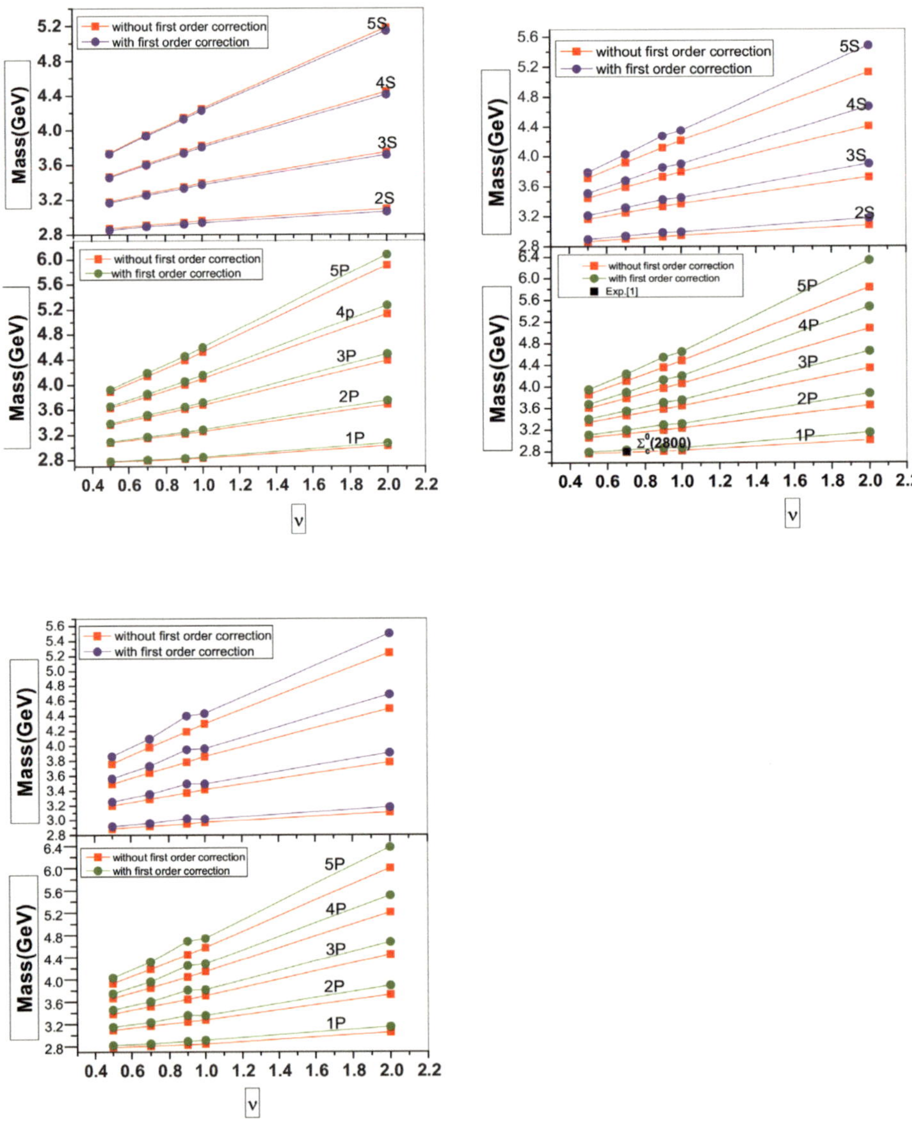

Figure 5.
Spectra of Σ triplets in S and P state with and without first order correction [13].

Recently, we calculate the masses of N^* and Δ resonances using the hCQM model by adding the first-order correction to the potential. The complete mass spectra with individual states graphically compared with the experimental states in **Figure 6**. We can observe that, many states are in accordance with the experimental resonances. We also predicted the J^P values of unknown states.

3.1 Regge trajectoy

We can say that, the mass spectra of hadrons can be conveniently described through Regge trajectories. These trajectories will aid in identifying the quantum number of particular resonance states. The important three properties of Regge Trajectories are: Linearity, Divergence and Parallelism.The higher excited radial and orbital states mass calculation enable to construct the Regge trajectories in the (n, M^2) and (J, M^2) planes. These graphical representation is useful in assigning J^P value to the experimental unkown states. We are getting almost linear, parallel and equidistance lines in each case of the baryons.

Some of the obtained masses are plotted in accordance with quantum number as well as with natural and unnatural parities. For the singly heavy baryons, the trajectory is shown for Ξ_c doublets baryons (See, **Figure 7**). For the doubly heavy baryons, the spectra of Ξ_{bc} and Ω_{bc} are less determined till the date. Thus, we plotted their trajetories in **Figure 8**. The natural and unnatural parities are shown in (J, M^2) for all triply heavy baryons; Ω_{ccc}, Ω_{bbb}, Ω_{bbc} and Ω_{ccb} (See, **Figures 9** and **10**). The rest trajetories of all heavy baryon families in both planes can be found in our articles.

3.2 Decays: strong, radiative and weak

The particles which are known to us decay by a similar sort of dissipation. Those who decay rapidly are unstable and take a long time are metastable. Some particles, like electron, three lightest neutrinos(and their antiparticles), the photon are stable

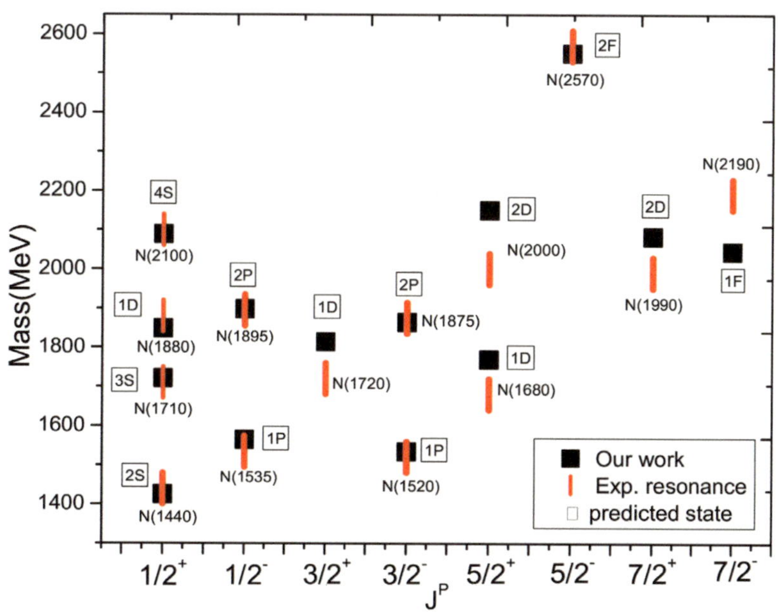

Figure 6.
The resonances are predicted from the first radial excited state (2S) to the orbital excited state (2F) for N^* baryons [19].

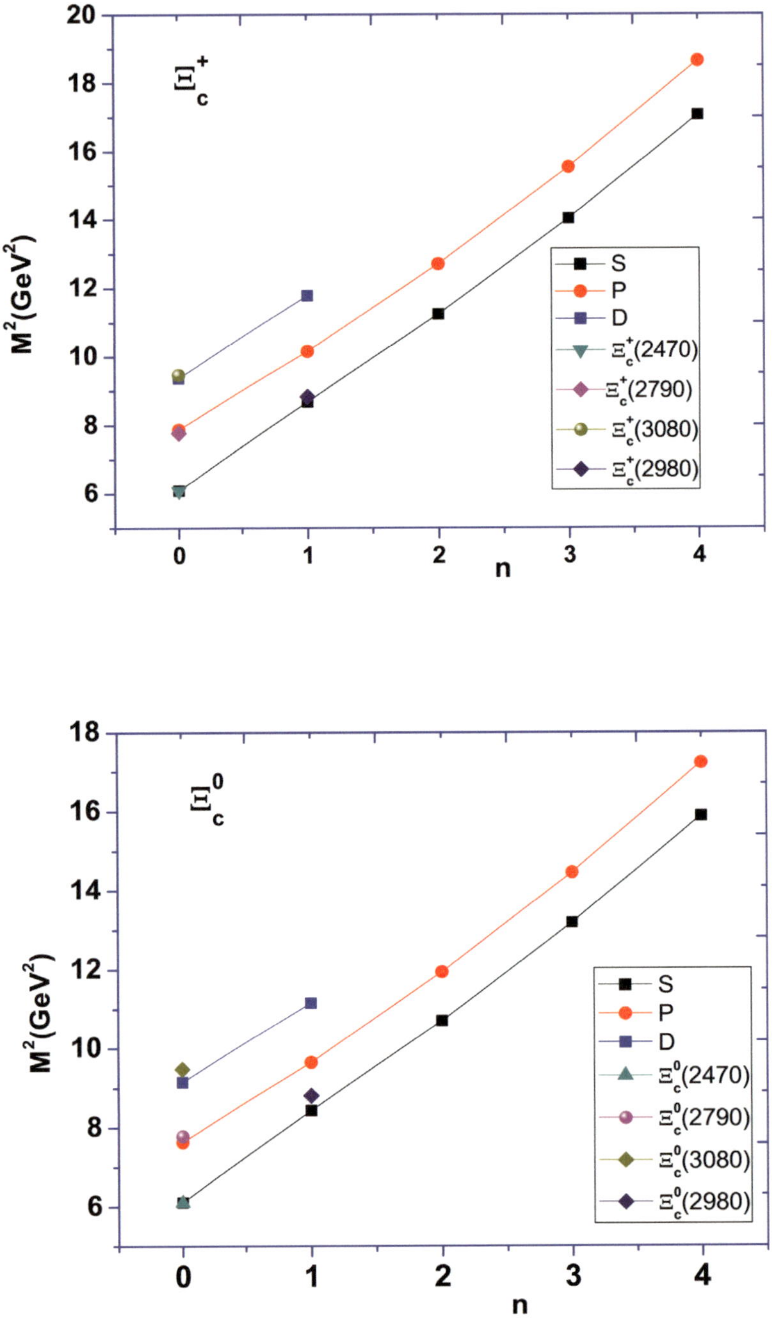

Figure 7.
Variation of mass for Ξ_c^0 and Ξ_c^+ with different states. The ($M^2 \to n$) Regge trajectories for J^P values $\frac{1^+}{2}$, $\frac{1^-}{2}$ and $\frac{5^+}{2}$ are shown from bottom to top. Available experimental data are also given with particle names [13].

partices (*never decays*). The observations of decays of baryon and meson resonances afford a valuable guidance in assigning to the correct places in various symmetry schemes. The correct isotopic spin assignment is likely to be implied by the experimental branching ratio into different charge states of particles produced by the decay,while experimental decay widths provide a means of extracting phenomenological coupling constants. For the success of a particular model, it is required to produce not only the mass spectra but also the decay properties of these baryons.

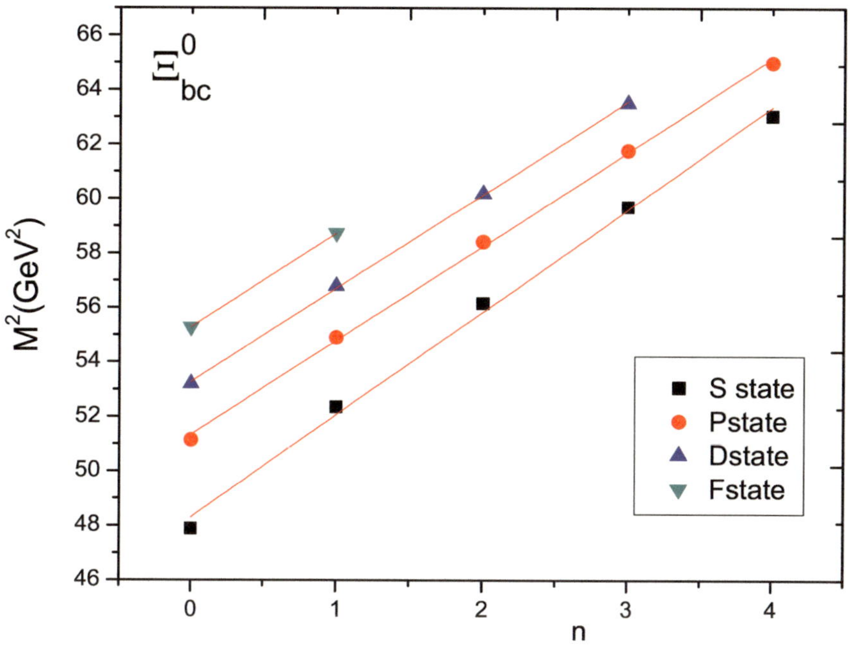

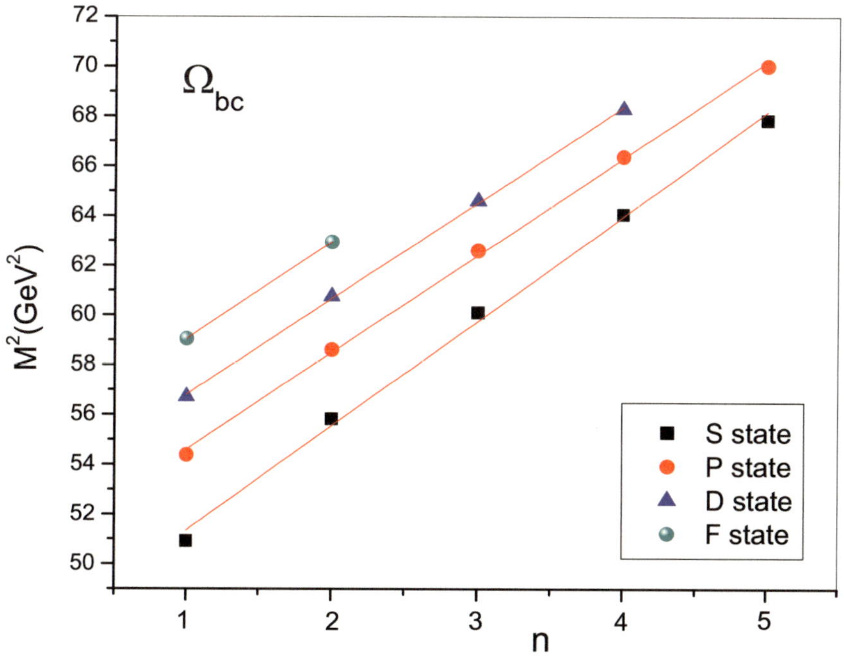

Figure 8.
The doubly charm-beauty baryons in ($M^2 \rightarrow n$) plane [14, 15].

For the success of a particular model, it is required to produce not only the mass spectra but also the decay properties of these baryons. The masses obtained from the hypercentral Constitute Quark Model (hCQM) are used to calculate the radiative and the strong decay widths. Such calculated widths are reasonably close to

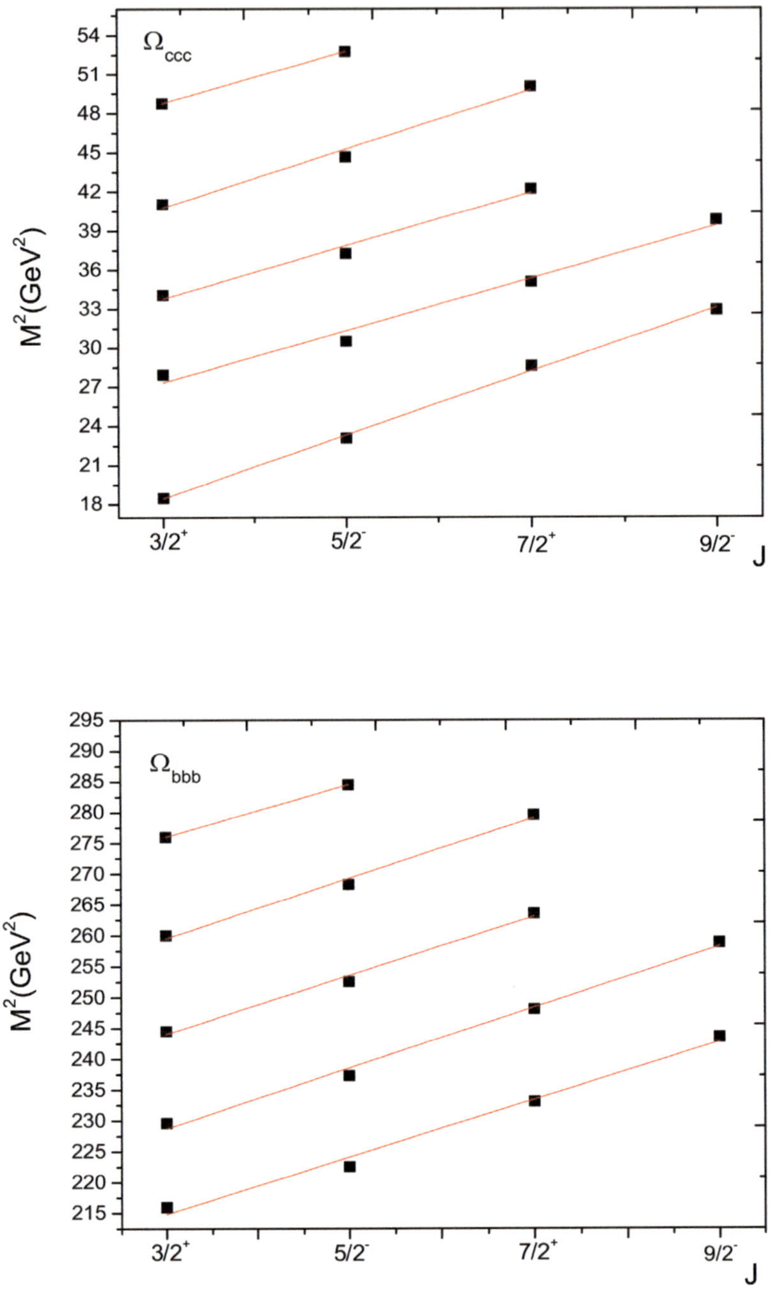

Figure 9.
The triply charm and triply bottom Regge trajectories in (J, M^2) plane [16].

other model predictions and experimental observations (where available) [30]. The effective coupling constant of the heavy baryons is small, which leads to their strong interactions perturbatively and makes it easier to understand the systems containing only light quarks. The Heavy Hadron Chiral Perturbation Theory (HHCPT) describes the strong interactions in the low-energy regime by an exchange of light Goldstone boson. Some of the strong P -wave couplings among the s-wave baryons and S-wave couplings between the s-wave and p-wave baryons are shown in **Table 2** with PDG values.

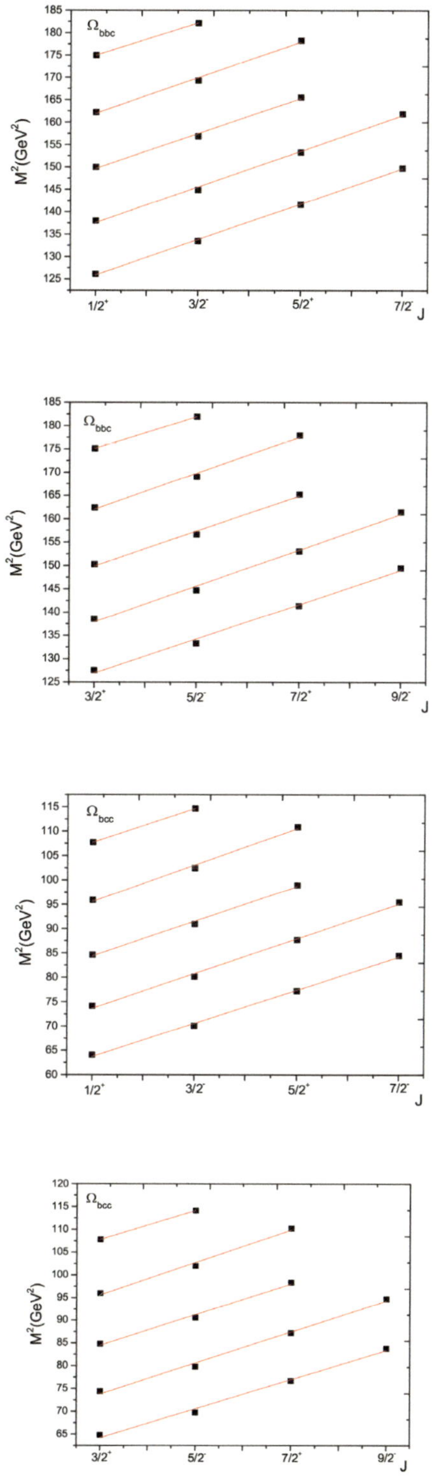

Figure 10.
Parent and daughter (J, M^2) Regge trajectories for triply heavy charm-beauty baryons with natural (first) and unnatural (second) parities [17, 18].

The electromagnetic properties are one of the essential key tools in understanding the internal structure and geometric shapes of hadrons. In the present study, the magnetic moments of heavy flavor and light flavor baryons are computed based on

	Decay mode	Present	PDG
P–wave transitions	$\Sigma_c^{++}\left(1^2S_{\frac{1}{2}}\right) \to \Lambda_c^+\pi^+$	1.72	$1.89^{+0.09}_{-0.18}$
P–wave transitions	$\Sigma_c^{+}\left(1^2S_{\frac{1}{2}}\right) \to \Lambda_c^+\pi^0$	1.60	< 4.6
P–wave transitions	$\Sigma_c^{0}\left(1^2S_{\frac{1}{2}}\right) \to \Lambda_c^+\pi^-$	1.17	$1.83^{+0.11}_{-0.19}$
S–wave transitions	$\Sigma_c^{++}\left(1^2P_{\frac{1}{2}}\right) \to \Lambda_c^+\pi^+$	68.19	75^{+22}_{-17}
S–wave transitions	$\Sigma_c^{+}\left(1^2P_{\frac{1}{2}}\right) \to \Lambda_c^+\pi^0$	62.92	62^{+60}_{-40}
S–wave transitions	$\Sigma_c^{0}\left(1^2P_{\frac{1}{2}}\right) \to \Lambda_c^+\pi^-$	66.44	72^{+22}_{-15}
S–wave transitions	$\Lambda_c^{+}\left(1^2P_{\frac{1}{2}}\right) \to \Sigma_c^0\pi^+$	4.45	2.6 ± 0.6

Table 2.
Several strong one-pion decay rates (in MeV) [30].

the non-relativistic hypercentral constituent quark model using the spin-flavor wave functions of the constituent quark and their effective masses [11, 13]. Generally, the meaning of the constituent quark mass corresponds to the energy that the quarks have inside the color singlet hadrons, we call it as the effective mass. The magnetic moment of baryons are obtained in terms of the spin, charge and effective mass of the bound quarks. The study has been performed for all singly, doubly and triply heavy baryon systems for positive parity $J^P = \frac{1}{2}^+, \frac{3}{2}^+$.

$$\mu_B = \sum_i \left\langle \phi_{sf} | \mu_{iz} | \phi_{sf} \right\rangle \tag{9}$$

where

$$\mu_i = \frac{e_i \sigma_i}{2m_i^{eff}} \tag{10}$$

e_i is a charge and σ_i is the spin of the respective constituent quark corresponds to the spin flavor wave function of the baryonic state. The effective mass for each of the constituting quark m_i^{eff} can be defined as [31].

$$m_i^{eff} = m_i \left(1 + \frac{\langle H \rangle}{\sum_i m_i} \right) \tag{11}$$

where, $\langle H \rangle$ = E + $\langle V_{spin} \rangle$. Using these equations, we calculate magnetic moments of singly, doubly and triply heavy baryons. The spin flavor wave function [32] ϕ_{sf} of all computed heavy flavor baryons are given in **Table 3**.

The electromagnetic radiative decay width is mainly the function of radiative transition magnetic moment $\mu_{B'_C \to B_c}$ (in μ_N) and photon energy (k) [12] as

$$\Gamma_\gamma = \frac{k^3}{4\pi} \frac{2}{2J+1} \frac{e}{m_p^2} \mu^2_{B_c \to B'_c} \tag{12}$$

where m_p is the mass of proton, J is the total angular momentum of the initial baryon (B_c). Some radiative decays are mentioned below:

- $\Sigma_c^{*0} \rightarrow \Sigma_c^0$: 1.553
- $\Xi_c^{*} \rightarrow \Xi_c^0$: 0.906
- $\Omega_c^{*0} \rightarrow \Omega_c^0$: 1.441
- $\Sigma_c^{*0} \rightarrow \Lambda_c^+$: 213.3

Weak decays of heavy hadrons play a crucial role to understand the heavy quark physics. In these decays the heavy quark acts as a spectator and the light quark inside heavy hadron decays in weak interaction [33]. The transition can be $s \rightarrow u$ or $d \rightarrow u$ depending on the available phase space. Since the heavy quark is spectator in such case, one can investigate the behavior of light quark system. These kind of small phase space transition could be possible in semi-electronic, semi-muonic and

Baryons	function	Our	Baryons	function	Our
$n(udd)$	$\frac{4}{3}\mu_d - \frac{1}{3}\mu_u$	−1.997	Δ	$2\mu_u + \mu_d$	2.28
$\Sigma_c^{++}(uuc)$	$\frac{4}{3}\mu_u - \frac{1}{3}\mu_c$	1.834	Σ_c^{*++}	$2\mu_u + \mu_c$	3.263
$\Sigma_c^0(udc)$	$\frac{4}{3}\mu_d - \frac{1}{3}\mu_c$	−1.091	Σ_c^{*+}	$\mu_u + \mu_d + \mu_c$	1.1359
$\Sigma_c^+(ddc)$	$\frac{2}{3}\mu_u + \frac{2}{3}\mu_d - \frac{1}{3}\mu_c$	0.690	Σ_c^{*0}	$2\mu_d + \mu_c$	−1.017
$\Xi_c^0(dsc)$	$\frac{2}{3}\mu_d + \frac{2}{3}\mu_s - \frac{1}{3}\mu_c$	−1.011	Ξ_c^{*0}	$\mu_d + \mu_s + \mu_c$	−0.825
$\Xi_c^+(usc)$	$\frac{2}{3}\mu_u + \frac{2}{3}\mu_s - \frac{1}{3}\mu_c$	0.559	Ξ_c^{*+}	$\mu_u + \mu_s + \mu_c$	1.329
$\Omega_c^0(ssc)$	$\frac{4}{3}\mu_s - \frac{1}{3}\mu_c$	−0.842	Ω_c^{*0}	$2\,\mu_s + \mu_c$	−0.625
$\Sigma_b^+(uub)$	$\frac{4}{3}\mu_u - \frac{1}{3}\mu_b$	2.288	Σ_b^{*+}	$2\mu_u + \mu_b$	3.343
$\Sigma_b^-(ddb)$	$\frac{4}{3}\mu_d - \frac{1}{3}\mu_b$	−1.079	Σ_b^{*-}	$2\mu_d + \mu_b$	−1.709
$\Xi_b^0(usb)$	$\frac{2}{3}\mu_u + \frac{2}{3}\mu_s - \frac{1}{3}\mu_b$	0.798	Ξ_b^{*0}	$\mu_u + \mu_s + \mu_b$	1.072
$\Xi_b^-(dsb)$	$\frac{2}{3}\mu_d + \frac{2}{3}\mu_s - \frac{1}{3}\mu_b$	−0.943	Ξ_b^{*-}	$\mu_d + \mu_s + \mu_b$	−1.471
$\Omega_b^-(ssb)$	$\frac{4}{3}\mu_s - \frac{1}{3}\mu_b$	0.761	Ω_b^{*-}	$2\,\mu_s + \mu_b$	−1.228
$\Xi_{cc}^+(dcc)$	$\frac{4}{3}\mu_c - \frac{1}{3}\mu_d$	0.784	Ξ_{cc}^{*+}	$2\,\mu_c + \mu_d$	0.068
$\Xi_{cc}^{++}(ucc)$	$\frac{4}{3}\mu_c - \frac{1}{3}\mu_u$	0.031	Ξ_{cc}^{*++}	$2\,\mu_c + \mu_u$	2.218
$\Xi_{bb}^-(dbb)$	$\frac{4}{3}\mu_b - \frac{1}{3}\mu_d$	0.196	Ξ_{bb}^{*-}	$2\,\mu_b + \mu_d$	−1.737
$\Xi_{bb}^0(ubb)$	$\frac{4}{3}\mu_b - \frac{1}{3}\mu_u$	−0.662	Ξ_{bb}^{*0}	$2\,\mu_b + \mu_u$	1.6071
$\Xi_{bc}^0(dbc)$	$\frac{2}{3}\mu_b + \frac{2}{3}\mu_c - \frac{1}{3}\mu_d$	0.527	Ξ_{bc}^{*0}	$\mu_b + \mu_c + \mu_d$	−0.448
$\Xi_{bc}^+(ubc)$	$\frac{2}{3}\mu_b + \frac{2}{3}\mu_c - \frac{1}{3}\mu_u$	−0.304	Ξ_{bc}^{*+}	$\mu_b + \mu_c + \mu_u$	2.107
$\Omega_{cc}^+(ccs)$	$\frac{4}{3}\mu_c - \frac{1}{3}\mu_s$	0.692	Ω_{cc}^+	$2\,\mu_c + \mu_s$	0.285
Ω_{bb}^-(bbs)	$\frac{4}{3}\mu_b - \frac{1}{3}\mu_s$	0.108	Ω_{bb}^{*-}	$2\,\mu_b + \mu_s$	−1.239
Ω_{bc}^0(bcs)	$\frac{2}{3}\mu_b + \frac{2}{3}\mu_c - \frac{1}{3}\mu_s$	0.439	Ω_{bc}^{*0}	$\mu_b + \mu_c + \mu_s$	−0.181
$\Omega_{bcc}^+(bcc)$	$\frac{4}{3}\mu_c - \frac{1}{3}\mu_b$	0.606	Ω_{bcc}^{*+}	$\mu_b + 2\mu_c$	0.8198
$\Omega_{bbc}^0(bbc)$	$\frac{4}{3}\mu_b - \frac{1}{3}\mu_c$	−0.233	Ω_{bbc}^{*+}	$2\mu_b + \mu_c$	0.228

Table 3.
Magnetic moments (in nuclear magnetons) with spin-flavor wavefunctions of $J^P = \frac{1}{2}^+, \frac{3}{2}^+$ *are listed for nucleons (light) and singly, doubly, triply(heavy) baryons.*

Mode	$J^P \to J'^{P'}$	Decay Rates(GeV)		Ref. [33]
$\Xi_c^0 \to \Lambda_c^+ e^- \bar{\nu}$	$\frac{1}{2}^+ \to \frac{1}{2}^+$	7.839×10^{-19}	7.839×10^{-19}	7.91×10^{-19}
$\Xi_c^0 \to \Sigma_c^+ e^- \bar{\nu}$	$\frac{1}{2}^+ \to \frac{1}{2}^+$	4.416×10^{-23}	7.023×10^{-24}	6.97×10^{-24}
$\Omega_c^0 \to \Xi_c^+ e^- \bar{\nu}$	$\frac{1}{2}^+ \to \frac{1}{2}^+$	2.143×10^{-18}	2.290×10^{-18}	2.26×10^{-18}
$\Omega_c^0 \to \Xi_c^{*+} e^- \bar{\nu}$	$\frac{1}{2}^+ \to \frac{3}{2}^+$	2.057×10^{-28}	2.436×10^{-28}	1.49×10^{-29}
$\Xi_b^- \to \Lambda_b^0 e^- \bar{\nu}$	$\frac{1}{2}^+ \to \frac{1}{2}^+$	5.928×10^{-19}		6.16×10^{-19}
$\Omega_b^- \to \Xi_b^0 e^- \bar{\nu}$	$\frac{1}{2}^+ \to \frac{1}{2}^+$	4.007×10^{-18}		4.05×10^{-18}
$\Omega_b^- \to \Xi_b^{*0} e^- \bar{\nu}$	$\frac{1}{2}^+ \to \frac{3}{2}^+$	1.675×10^{-26}		3.27×10^{-28}

Table 4.
Semi-electronic decays in $s \to u$ transition for charm baryons are listed [13].

non leptonic decays of the heavy baryons and mesons. We calculate here, the semi-electronic decays for strange-charm heavy flavor baryons Ω_c, Ξ_c, Ξ_b and Ξ_b using our spectral parameters.

The differential decay rates for exclusive semi-electronic decays are given by [33],

$$\frac{d\Gamma}{dw} = \frac{G_F^2 M^5 |V_{CKM}|^2}{192\pi^3} \sqrt{w^2 - 1} P(w) \tag{13}$$

where $P(w)$ contains the hadronic and leptonic tensor. Assuming that the form factors are slowly varying functions of the kinematic variables, we may replace all form factors by their values at variable w=1. The calculated semi-electronic decays for Ξ_c, Ω_c, Ξ_b and Ω_b baryons are listed in **Table 4**. We can observe that our results are in accorance with ref. [33] for singly heavy baryons.

4. Current cenario in the field of baryons

Baryons with heavy quarks provide a beautiful laboratory to test our ideas of QCD. As the heavy quarks mass increases its motion decreases and the baryons properties are increasingly governed by the dynamics of the light quark and approach a universal limit. As we discussed, many theoretically approaches are calculating and obtaning the masses and decay widths of heavy baryons. We study the mass spectroscopy of ligt and heavy baryons and their properties. A few number of excited states for the singly heavy baryons have also been reported along with their ground states. The singly charmed baryons are, $\Lambda_c(2286)^+$, $\Lambda_c(2595)^+$, $\Lambda_c(2625)^+$, $\Lambda_c(2880)^+$, $\Lambda_c(2940)^+$, $\Lambda_c(2765)^+$,$\Lambda_c(2860)^+$, $\Sigma_c(2455)^{++,+,0}$, $\Sigma_c(2520)^{++,+,0}$, $\Sigma_c(2800)^{++,+,0}$,$\Xi_c(2468)^{+,0}$, $\Xi_c'(2580)^{+,0}$, $\Xi_c'(2645)^{+,0}$, $\Xi_c(2790)^{+,0}$, $\Xi_c(2815)^{+,0}$, $\Xi_c'(2930)^0$, $\Xi_c(2980)^{+,0}$, $\Xi_c(3055)^+$, $\Xi_c(3080)^{+,0}$, $\Xi_c'(3123)^+$, $\Omega_c(2695)^0$, $\Omega_c(2770)^0$. The singly beauty baryons are $\Lambda_b(5619)^0$, $\Lambda_b(5912)^0$, $\Lambda_b(5920)^0$, $\Sigma_b(5811)^+$, $\Sigma_b(5816)^-$, $\Sigma_b(5832)^+$, $\Sigma_b(5835)^-$, $\Xi_b(5790)^{-,0}$, $\Xi_b(5945)^0$, $\Xi_b(5955)^-$, $\Xi_b(5935)'^-$. And now, LHCb experiment has identified new resonances(excited states) for singly heavy baryons (See, **Table 5**). The experimental state and masses are in first two columns. The third column shows our predicted masees and in the forth column we assign the states with J^P values. We can observe that, apart from first radial and orbital excited states, we also have D state resonances in heavy

Names	Exp. Mass (MeV) [2]	Predicted Mass (MeV)	Baryon State
$\Lambda_c^+(2860)$	2756.1 ± 0.5	2842	$\left(1^2D_{\frac{3}{2}}\right)$
$\Omega_c(3000)^0$	$3000.4 \pm 0.2 \pm 0.1^{0.3}_{-0.5}$	2976,2993	$\left(1^2P_{\frac{3}{2}}\right),\left(1^4P_{\frac{3}{2}}\right)$
$\Omega_c(3050)^0$	$3050.4 \pm 0.1 \pm 0.1^{0.3}_{-0.5}$	3011,3028	$\left(1^2P_{\frac{1}{2}}\right),\left(1^4P_{\frac{1}{2}}\right)$
$\Omega_c(3066)^0$	$3065.6 \pm 0.1 \pm 0.3^{0.3}_{-0.5}$	2947	$\left(1^4P_{\frac{5}{2}}\right)$
$\Omega_c(3090)^0$	$3090.2 \pm 0.3 \pm 0.5^{0.3}_{-0.5}$	3011,3028	$\left(1^2P_{\frac{1}{2}}\right),\left(1^4P_{\frac{1}{2}}\right)$
$\Omega_c(3119)^0$	$3119.1 \pm 0.3 \pm 0.9^{0.3}_{-0.5}$	3100,3126	$\left(2^2S_{\frac{1}{2}}\right),\left(2^2S_{\frac{1}{2}}\right)$
$\Xi_{cc}^{++}(3620)$	3521.4 ± 0.99	3511	$\left(1^2S_{\frac{1}{2}}\right)$
$\Xi_c^0(2965)$	2964.88 ± 0.26	2903,2940	$\left(1^2S_{\frac{1}{2}}\right)$
$\Lambda_b^+(6072)$	6072.3 ± 2.9	6066	$\left(2^2S_{\frac{1}{2}}\right)$
$\Lambda_b^+(6146)$	6146.17 ± 0.33	6121	$\left(1^2D_{\frac{3}{2}}\right)$
$\Lambda_b^+(6152)$	6152.51 ± 0.26	6119	$\left(1^2D_{\frac{5}{2}}\right)$
$\Sigma_b^+(6097)^-$	6098 ± 1.7	6122	$\left(1^2P_{\frac{1}{2}}\right)$
$\Sigma_b(6097)^+$	6095.8 ± 1.7	6131	$\left(1^2P_{\frac{1}{2}}\right)$
$\Omega_b(6316)^-$	$6315.64 \pm 0.31 \pm 0.07$	6313	$\left(1^4P_{\frac{5}{2}}\right)$
$\Omega_b(6330)^-$	$6330.3 \pm 0.28 \pm 0.07$	6331	$\left(1^2P_{\frac{1}{2}}\right),\left(1^4P_{\frac{1}{2}}\right)$
$\Omega_b(6340)^-$	$6339.71 \pm 0.26 \pm 0.05$	6321	$\left(1^2P_{\frac{3}{2}}\right)$
$\Omega_b(6350)^-$	$6349.88 \pm 0.35 \pm 0.05$	6326	$\left(1^4P_{\frac{3}{2}}\right)$
$\Xi_b(6227)^0$	$6227.1^{1.4}_{1.5} \pm 0.5$	6193,6309	$\left(2^2S_{\frac{1}{2}}\right),\left(2^4S_{\frac{3}{2}}\right)$

Table 5.
The newly observed baryonic states are listed with the observed mass in column 1 & 2. Our predicted baryonic states are compared with J^P value and masses.

sector. According to the SU(3) symmetry we also have doubly and triply baryons in charm as well as bottom sector. Among which, the evidence of 1S state for doubly heavy baryons Ξ_{cc}^+ and Ξ_{cc}^{++} are observed by the SELEX and the LHCb experiments respectively. The Belle and CDF collaboration had also observed some of the particles. The fulture experiments like Panda, Belle-II and BES-II are expected to give more results soon.

Author details

Zalak Shah* and Ajay Kumar Rai
Department of Physics, Sardar Vallabhbhai National Institute of Technology, Surat, Gujarat, India

*Address all correspondence to: zalak.physics@gmail.com

References

[1] Zyla P.A. et al., (Particle Data Group), Prog. Theor. Exp. Phys. 2020; 083C01:2020.

[2] Aaij, Roel et al; JHEP 2017; 05:030; Phys. Rev. Lett. 2017; 18:182001; Phys. Rev. Lett.(sigmab) 122, 012001 (2019); Phys. Rev. (lambdab)Lett. 123, 152001 (2019);Phys. Rev. Lett. (omgb)124, 082002 (2020); JHEP 06,(lamb) 136 (2020); arXiv:2010.14485v1(xib) (2020); Phys. Rev. Lett.(xic) 124, 222001 (2020)

[3] Capstick S, Isgur N, Baryons in a relativized quark model with chromodynamics, Phys. Rev. D 1986; 34:2809

[4] Klempt E, Richard J M, Baryon spectroscopy, Rev. Mod. Phys. 2010; 82: 1095.

[5] Crede, V. and Roberts, W, Progress towards understanding baryon resonances, Rept. Prog. Phys. 2013; 76: 076301

[6] Ebert D, Faustov R N, Galkin V O, Spectroscopy and Regge trajectories of heavy baryons in the relativistic quark-diquark picture, Phys. Rev. D 2011;84: 014025.

[7] Valcarce A, Garcilazo H, Vijande J, Towards an understanding of heavy baryon spectroscopy, Eur. Phys. J. A 2008; 37:217.

[8] Padmanath M, Edwards Robert G and Mathur N Peardon M, Excited-state spectroscopy of singly, doubly and triply-charmed baryons from lattice QCD, Proceedings, 6th International Workshop on Charm Physics (Charm 2013) 2013; arxiv:1311.4806.

[9] Gandhi K, Shah Z and Rai A K, Spectrum of nonstrange singly charmed baryons in the constituent quark model, Int J Theor Phys 2020 59: 11291156.

[10] Shah Z and Rai A K, Excited mass spectra of Ω_b Baryon Few Body Syst. 2018; 59:112.

[11] Thakkar K, Shah Z, Rai A K and Vinodkumar P C, Excited State Mass spectra and Regge trajectories of Bottom Baryons, Nucl. Phys. A 2017; 965:57.

[12] Shah Z,Thakkar K and Rai A K and Vinodkumar P C, Mass spectra and Regge trajectories of Λ_c^0, Σ_c^0, Ξ_c^0, Ω_c^0 Baryons, Chin. Phys. C. 2016; 40: 123102.

[13] Shah Z, Thakkar K, Rai A K and Vinodkumar P C, Excited State Mass spectra of Singly Charmed Baryons, Eur. Phys. J. A 2016; 52:313.

[14] Shah Z, Thakkar K and Rai A K, Excited State Mass spectra of Doubly heavy Baryons $\Omega_c c$, $\Omega_b b$ and $\Omega_b c$, Eur. Phys. J. C 2016; 76:530.

[15] Shah Z and Rai A K, Excited State Mass spectra of doubly heavy Ξ baryons, Eur. Phys. J. C 2017; 77:129.

[16] Shah Z and Rai A K Masses and Regge trajectories of triply heavy Ω_{ccc} and Ω_{bbb} baryons, Eur. Phys. J. A 2017, 53:195.

[17] Shah Z and Rai A K, Spectroscopy of Ω_{ccb} baryon in Hypercentral quark Model, Chin. Phys. C. 2018, 42:053101.

[18] Shah Z and Rai A K, Ground and excited state masses of Ω_{bbc} baryon, Few Body Syst. 2018; 59:76.

[19] Shah Z, Gandhi K, Rai A K, Spectroscopy of light N^* baryons, Chin. Phys. C 2019; 43:024106

[20] Menapara C, Shah Z, Rai A K 2020; *arXiv:2010.04386*

[21] Gell-Mann M, A Schematic of Baryons and Mesons, Phys. Lett.1964; 8 (3):214–215

[22] Griffiths, David J.INTRODUCTION TO EL, EMENTARY PARTICLES 1987 NEW YORK, USA: WILEY (1987) 392p

[23] Aaij R et al.,[LHCb Collaboration] Observation of $J/\psi p$ Resonances Consistent with Pentaquark States in $\Lambda_b^0 \rightarrow J/\psi K^- p$ Decays, Phys. Rev. Lett. 2015; 115:072001

[24] Azizi K, Sarac Y, Sundu H, Properties of $P_c^+(4312)$ pentaquark arxiv:2011.05828 2020.

[25] Jaffe R L, Perhaps a Stable Dihyperon, Phys. Rev. Lett. 1977; 38: 195

[26] Rai A K, Pandya J N, Vinodkumar P C, Indian J. Phys. A 2006, 80:387–392.

[27] Rai A K, Pandya J N, Vinodkumar P C, Nucl. Phys. A 2007, 782:406–409

[28] Bijker R, Iachello F, and Leviatan A, Algebraic models of hadron structure Nonstrange baryons, Annals Phys. 1994; 236: 69–116; Annals Phys. 2000: 284:89.

[29] Koma Y and Koma M and Wittig H, Nonperturbative determination of the QCD pote;tial at O(1/m), Phys. Rev. Lett. 2006; 97:122003

[30] Gandhi K, Shah Z and Rai A K; Decay properties of singly charmed baryons, Eur. Phys. J. Plus 2018; 133:512.

[31] Silvestre-Brac, B, Spectroscopy of baryons containing heavy quarks, Prog. Part. Nucl. Phys. 1996; 36: 263–273

[32] Fayyazuddin and Riazuddin, A Modern introduction to particle physics, 1994.

[33] Faller S and Mannel T, Light-Quark Decays in Heavy Hadrons, Phys. Lett. 2015; B750:653–659.

Section 3

QCD Topics

Chapter 5

Double Pole Method in QCD Sum Rules for Vector Mesons

Mikael Souto Maior de Sousa and Rômulo Rodrigues da Silva

Abstract

The QCD Sum Rules approach had proposed by Shifman, Vaishtein Zakharov Novikov, Okun and Voloshin (SVZNOV) in 1979 and has been used as a method for extracting useful properties of hadrons having the lowest mass, called as ground states. On the other hand, the most recent experimental results make it clear that the study of the excited states can help to solve many puzzles about the new XYZ mesons structure. In this paper, we propose a new method to study the first excited state of the vector mesons, in particular we focus our attention on the study of the ρ vector mesons, that have been studied previously by SVZNOV method. In principle, the method that we used is a simple modification to the shape of the spectral density of the SVZNOV method, which is written as "pole + continuum", to a new functional form "pole + pole + continuum". In this way, We may obtain the ρ and the $\rho(2S)$ masses and also their decay constants.

Keywords: QCD Sum Rules, double pole, light quarks, vector mesons

1. Introduction

The successful QCD sum rules was created in 1977 by Shifman, Vainshtein, Zakharov, Novikov, Okun and Voloshin [1–4], and until today is widely used. Using this method, we may obtain many hadron parameters such as: hadrons masses, decay constant, coupling constant and form factors, all they giving in terms of the QCD parameters, it means, in terms of the quark masses, the strong coupling and nonperturbative parameters like quark condensate and gluon condensate.

The main point of this method is that the quantum numbers and content of quarks in hadron are presented by an interpolating current. So, to determine the mass and the decay constant of the ground state of the hadron, we use the two-point correlation function, where this correlation function is introduced in two different interpretations. The first one is the OPE's interpretation, where the correlation function is presented in terms of the operator product expansion (OPE).

On the phenomenological side we can be written the correlation function in terms of the ground state and several excited states. The usual QCDSR method uses an ansatz that the phenomenological spectral density can be represented by a form "pole + continuum", where it is assumed that the phenomenological and OPE spectral density coincides with each other above the continuum threshold. The continuum is represented by an extra parameter called $s0$, as being correlated with the onset of excited states [5].

In general, the resonances occurs with $\sqrt{s_0}$ lower than the mass of the first excited state. For the ρ meson spectrum, for example, the ansatz "pole + continuum" is a good approach, due to the large decay width of the $\rho(2S)$ or $\rho(1450)$, which allows to approximate the excited states as a continuum. For the ρ meson [6], the value of $\sqrt{s_0}$ that best fit the mass and the decay constant is $\sqrt{s_0} = 1.2\ GeV$ and for the $\varphi(1020)$ meson the value is 1.41 GeV. We note that the values quoted above for $\sqrt{s_0}$ are about 250 MeV below the poles of $\rho(1450)$ and $\varphi(1680)$. One interpretation of this result is due to the effect of the large decay width of these mesons.

Novikov et al. [1], in a pioneering paper, proposed, for the charmonium sum rule, that the phenomenological side with double pole ("pole + pole + continuum") and $\sqrt{s'_0} = 4\ GeV$, where s'_0 is the new parameter that takes in to account the second "pole" in this ansatz. In this way, thi value is correlated with the threshold of pair production of charmed mesons. Using this $\sqrt{s'_0}$ value and the Sum Rule Momentum at $Q^2 = 0$, they presented the first estimate for the gluon condensate and a very good estimated value for η_c meson, that is about 3 GeV, while the experimental data had shown 2.83 GeV for this meson [1, 2, 4].

In general, by QCDSR, the excited states are studied in "pole + pole + continuum" ansatz with $Q^2 = 0$ [1, 2], as we can see in the spectral sum [7], the Maximum Entropy [8] and Gaussian Sum Rule with "pole + pole + continuum" ansatz [9] approaches. There are studies on the $\rho(1S, 2S)$ mesons [8, 10, 11], nucleons [7, 12], $\eta c(1S, 2S)$ mesons [2], $\psi(1S, 2S)$ mesons [1, 13] and $\Upsilon(1S, 2S)$ mesons [14]. In this paper, we obtain the $\rho(1S, 2S)$ mesons masses and their decay constants taking the "pole + pole + continuum" ansatz in QCD sum rules,

2. The two point correlation function

As it is known, to determine the hadron mass and the decay constant in QCD sum rules, we may use the two-point correlation function [3], that is given by

$$\Pi_{\mu\nu} = i\int d^4x e^{iqx}\left\langle 0|T\{j_\mu(x)j^\dagger_\nu(0)|0\right\rangle, \tag{1}$$

where, on the OPE side, this current density for $q\bar{q}$ vector mesons has the following form:

$$j_\mu(x) = \delta_{ab}\bar{q}_a(x)\gamma_\mu q_b(x), \tag{2}$$

where the subscribe index a and b represents the color index. Now, using Eq. (2) in Eq. (1) we have

$$\Pi_{\mu\nu} = i\delta_{ab}\delta_{cd}\int d^4x e^{iqx}\left\langle 0|T\{\bar{q}_a(x)\gamma_\mu q_b(x)\left[\bar{q}_c(0)\gamma_\nu q_d(0)\right]^\dagger|0\right\rangle. \tag{3}$$

Evaluating Eq. (3) in terms of the OPE [15], which can be written by a dispersion relation, where this relation depends on the QCD parameters, the correlation function takes the form:

$$\Pi^{OPE}_{\mu\nu}(q) = \left(q_\mu q_\nu - q^2 g_{\mu\nu}\right)\Pi^{OPE}(q^2), \tag{4}$$

with:

$$\Pi^{OPE}(q^2) = \int_{s_0^{min}}^{\infty} ds\frac{\rho^{OPE}(s)}{s-q^2} + \Pi^{nonPert}(q^2), \tag{5}$$

Note that

$$\rho^{Pert}(s) = \frac{\mathrm{Im}\left[\Pi^{OPE}(s)\right]}{\pi}, \tag{6}$$

and $\Pi^{nonPert}(q^2)$ is the term that add the condensates contributions, besides, s_0^{min} is the minimum value of the s parameter to have an imaginary part of the $\Pi^{Pert}(s)$.

On the phenomenological side, the interpolating current density may be written considering just hadronic freedom degrees, it means, inserting a complete set of intermediate states among the operator, where they are the creation and annihilation describes by the interpolating current density. In this way we can use the following operator algebra

$$\left\langle 0|\, j_\mu(0)|V(q)\right\rangle = fmc_\mu(q), \tag{7}$$

where f and m are, respectively, the decay constant and the mass of the meson and $\epsilon_\mu(q)$ is a unitary polarizing vector. So, substituting Eq. (7) in Eq. (1) after some intermediate mathematical steps we get:

$$\Pi^{\mathrm{Phen}}_{\mu\nu}(q) = \left(q_\mu q_\nu - q^2 g_{\mu\nu}\right)\Pi^{Phen}\left(q^2\right). \tag{8}$$

The Invariant part of Eq. (8), $\Pi^{Phen}(q^2)$, is given by the following dispersion relation:

$$\Pi^{Phen}\left(q^2\right) = \int_{s_0^{min}}^{\infty} ds \frac{\rho^{Phen}(s)}{s - q^2}, \tag{9}$$

where $\rho^{Phen}(s) = f^2\delta(s - m^2) + \rho^{cont.}(s)$. Thus, we have the Eq. (9) written as follow:

$$\Pi^{Phen}\left(q^2\right) = \frac{f^2}{m^2 - q^2} + \int_{s_0}^{\infty} ds \frac{\rho^{cont.}(s)}{s - q^2}. \tag{10}$$

Note that, we can introduce a minimum number of parameters in the calculus by the approach $\rho^{cont.}(s) = \Theta(s - s_0)\rho^{OPE}(s)$, using this in Eq. (10) we get:

$$\Pi^{Phen}\left(q^2\right) = \frac{f^2}{m^2 - q^2} + \int_{s_0}^{\infty} ds \frac{\rho^{OPE}(s)}{s - q^2}, \tag{11}$$

so, s_0 can be understood as a parameter indicating that for s values greater than s_0 there is only contribution from the continuum, it means, s_0 is called a continuum threshold.

Note that, by the Quark-Hadron duality we can develop the two-point correlation function in both different interpretation that are equivalent each other. I.e., we can match the correlation function by de OPE, Eq. (5), with the correlation function by the Phenomenological side, Eq. (10), through the Borel transformation.

3. Borel transformation

To macht the Eqs. (5) and (10) is not that simple, because in the OPE side the calculations of the all OPE terms is almost impossible, in this way, at someone moment we must truncate the series and, beyond this, guaranties its convergence. However, for the truncation of the series to be possible, the contributions of the terms of higher dimensions must be small enough to justify to be disregarded in the expansion.

Thereby, for both descriptions to be in fact equivalent, we must suppress both the contributions of the highest order terms of the OPE and the contributions of the excited states on the phenomenological side. It is can be done by the Borel transformation that is define as follow:

$$B[\Pi(Q^2) = \Pi[M^2] = \lim_{\substack{Q^2, n \to \infty \\ \frac{Q^2}{n} = M^2}} \frac{(Q^2)^n}{n!} \left(-\frac{\partial}{\partial Q^2}\right)^n \Pi(Q^2), \tag{12}$$

where $Q^2 = -q^2$ is the momentum in the Euclidian space and M^2 is a variable rising due to Borel transformation application and it is called *Borel Mass*.

Because of this, we can determine a region of the M2 space in which both the highest order contributions from OPE and those from excited states are suppressed, so that the phenomenological parameters associated with the hadron fundamental state can be determined. Therefore, we must determine an interval of M2 where this comparison is adequate, enabling the determination of reliable results. This interval is called Borel Window.

At the Phenomenological side, we introduce some approximations when we assume that the spectral density can be considering as a polo plus a continuum of excited states. So, we must suppress the continuum contributions for the result to be sufficiently dominated by de pole.

4. The double pole method

This method is consisted by the assumption that the spectral density at the phenomenological side can be given like [16]:

$$\rho^{Phen}(s) = (f_1)^2 \delta\left[s - (m_1)^2\right] + (f_2)^2 \delta\left[s - (m_2)^2\right] + \rho^{OPE}(s)\Theta(s - s_0'), \tag{13}$$

where m_1 and f_1 are, respectively, the ground state meson mass and decay constant, m_2 and f_2 are, respectively, the first excited state meson mass and decay constant, beyond this, we include a new parameter s_0' marking the onset of the continuum states. As we can see in **Figure 1**, the parameters Δ and $\boldsymbol{\Delta}'$consists in a gap among the ground and first excited states and among the first excited and the continuum states respectively. They are defined by the decay width of these states.

Note that, inserting Eq. (13) in Eq. (9) we get the following two-point phenomenological correlation function:

$$\Pi^{Phen}(q^2) = \frac{(f_1)^2}{(m_1)^2 - q^2} + \frac{(f_2)^2}{(m_2)^2 - q^2} + \int_{s_0'}^{\infty} ds \frac{\rho^{OPE}(s)}{s - q^2}, \tag{14}$$

Applying the Borel transformation in Eqs. (14) and (5) we get:

$$\Pi^{Phen}(M^2) = (f_1)^2 e^{-(m_1)^2/M^2} + (f_2)^2 e^{-(m_2)^2/M^2} + \int_{s_0'}^{\infty} ds \rho^{OPE}(s) e^{-s/M^2}, \tag{15}$$

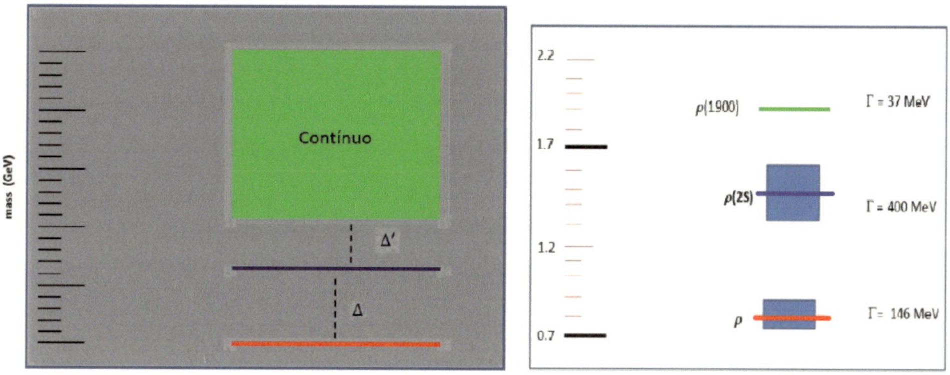

Figure 1.
*On the left side it is seen the double pole ansatz, **Δ** and **Δ'** represent the Gaps among the ground, first excited and continuum states. On the right side it is seen the mass spectra for the* **ρ** *meson and its resonances [16–18].*

$$\Pi^{OPE}\left(M^2\right) = \int_{s_0'^{min}}^{s_0'} ds\rho^{OPE}(s)e^{-s/M^2} + \int_{s_0'}^{\infty} ds\rho^{OPE}(s)e^{-s/M^2} + \Pi^{cond}\left(M^2\right). \tag{16}$$

By the Quark-Hadron duality we have $\Pi^{Phen}\left(M^2\right) = \Pi^{OPE}\left(M^2\right)$, thus:

$$\left(f_1\right)^2 e^{-(m_1)^2/M^2} + \left(f_2\right)^2 e^{-(m_2)^2/M^2} = \int_{s_0'^{min}}^{s_0'} ds\rho^{OPE}(s)e^{-s/M^2} + \Pi^{cond}\left(M^2\right). \tag{17}$$

The contribution of the resonances is given by:

$$CR = \int_{s_0'}^{\infty} ds\rho^{OPE}(s)e^{-s/M^2}. \tag{18}$$

To develop Eq. (17) let us make a variable change taking $M^{-2} = x$, so we write:

$$\left(f_1\right)^2 e^{-(m_1)^2 x} + \left(f_2\right)^2 e^{-(m_2)^2 x} = \int_{s_0'^{min}}^{s_0'} ds\rho^{OPE}(s)e^{-sx} + \Pi^{cond}(x). \tag{19}$$

Now, taking de derivative of Eq. (19) with respect to x we get:

$$-\left(m_1 f_1\right)^2 e^{-(m_1)^2/M^2} - \left(m_2 f_2\right)^2 e^{-\frac{(m_2)^2}{M^2}} = \frac{d}{dx}\Pi^{OPE}(x), \tag{20}$$

where, now, we are considering

$$\Pi^{OPE}(x) = \int_{s_0'^{min}}^{s_0'} ds\rho^{OPE}(s)e^{-sx} + \Pi^{cond}(x). \tag{21}$$

We observe that the Eqs. (19) and (21) form a equations system in x variable. In this system we can make a new change of variables as follow:

$$A(x) = \left(f_1\right)^2 e^{-(m_1)^2 x} \ \ and\ B(x) = \left(f_2\right)^2 e^{-(m_2)^2 x}, \tag{22}$$

this way we get the following system:

$$A(x) + B(x) = \Pi^{OPE}(x). \tag{23}$$

$$-(m_1)^2 A(x) - (m_2)^2 B(x) = \frac{d}{dx}\Pi^{OPE}(x) \tag{24}$$

Solving the above system of equation for $A(x)$ and $B(x)$ we have:

$$A(x) = \frac{\frac{d}{dx}\Pi^{OPE}(x) + \Pi^{OPE} m_2^2}{m_2^2 - m_1^2}. \tag{25}$$

$$B(x) = \frac{\frac{d}{dx}\Pi^{OPE}(x) + \Pi^{OPE} m_1^2}{m_1^2 - m_2^2}. \tag{26}$$

Note that, Eqs. (25) and (26) presents information about the hadron masses and their coupling constants, to eliminate de coupling constants we have to take the derivative of Eq. (25) and then dividing the result by the own Eq. (25) and the same procedure with Eq. (26). Thus, we have:

$$m_1 = \sqrt{-\frac{\frac{d}{dx}\Pi^{OPE}(x) m_2^2 + \frac{d^2}{dx^2}\Pi^{OPE}}{\frac{d}{dx}\Pi^{OPE}(x) + \Pi^{OPE} m_2^2}}, \tag{27}$$

$$m_2 = \sqrt{-\frac{\frac{d}{dx}\Pi^{OPE}(x) m_1^2 + \frac{d^2}{dx^2}\Pi^{OPE}}{\frac{d}{dx}\Pi^{OPE}(x) + \Pi^{OPE} m_1^2}}. \tag{28}$$

This way we have the both solutions coupling each other. On the other hand, what we are looking for are mass solutions for the ground state and its first excited state independent each other. To do so, we take the second derivative of the Eq. (19) with respect to x and the result we divide by Eq. (23) for decoupling of the m_1 mass. Note that the same procedure can be done for Eqs. (19) and (24) for decoupling of the m_2 mass. So, for the m_2 mass we have:

$$m_2^4 = \frac{\frac{d^3}{dx^3}\Pi^{OPE}(x) + m_1^2 \frac{d^2}{dx^2}\Pi^{OPE}(x)}{\frac{d}{dx}\Pi^{OPE}(x) + \Pi^{OPE}(x) m_1^2}. \tag{29}$$

Substituting Eq. (29) in Eq. (28) we obtain the following polynomial equation:

$$m_2^4 a + m_2^2 b + c = 0, \tag{30}$$

where a, b and c are respectively:

$$a = -\left(\frac{d}{dx}\Pi^{OPE}(x)\right)^2 + \Pi^{OPE}(x)\frac{d^2}{dx^2}\Pi^{OPE}(x), \tag{31}$$

$$b = -\left(\frac{d^2}{dx^2}\Pi^{OPE}(x)\right)\frac{d}{dx}\Pi^{OPE}(x) + \Pi^{OPE}(x)\frac{d^3}{dx^3}\Pi^{OPE}(x), \tag{32}$$

$$c = \left(\frac{d^3}{dx^3}\Pi^{OPE}(x)\right)\frac{d}{dx}\Pi^{OPE}(x) - \frac{d^2}{dx^2}\Pi^{OPE}(x). \tag{33}$$

Note that, for the m_1 mass, following the same procedure we get the other polynomial equation like Eq. (30). Thus, solving the polynomial equation, given by Eq. (30), and the same for m_1 mass, the physical solutions the represent m_1 like the ground state mass and m_2 like the first excited state mass are given by:

$$m_1 = \sqrt{-\frac{b + \sqrt{\Delta}}{2a}}, \tag{34}$$

$$m_2 = \sqrt{-\frac{b - \sqrt{\Delta}}{2a}}. \tag{35}$$

These results can be developed to obtain the masses of the ground state and its first excited state for any $\overline{q}q$ vector meson, also, we can calculate their coupling constants using the masses estimated in the Eqs. (25) and (26).

5. Results for the ρ meson

For the ρ meson we use the $\Pi^{OPE}(x)$ given by the Feynman diagrams that to be seen in [citar o greiber]. In this way we have:

$$\rho^{OPE}(s) = \frac{1}{4\pi^2}\left(1 + \frac{\alpha_s}{\pi}\right) \tag{36}$$

and

$$\Pi^{cond}(x) = x\left(\frac{1}{12}\left\langle\frac{\alpha_s}{\pi}G^2\right\rangle + 2\,m_q\langle\overline{q}q\rangle\right) - x^2\frac{112}{18}\pi\alpha_s\langle\overline{q}q\rangle^2, \tag{37}$$

where α_s is the strong coupling constant, m_q is the light quark mass, $\langle\frac{\alpha_s}{\pi}G^2\rangle$ is the gluon condensate, $\langle\overline{q}q\rangle$ is the quark condensate and $s_0^{min} = 4\,m_q^2$.

Following [19], for the ρ meson we use the parameters: $\alpha_s = 0.5$, $m_q = (6.4 \pm 1.25)\,MeV$, $\langle\overline{q}q\rangle = -(0.24 \pm 0.01)^3\,GeV^3$, $\left\langle\frac{\alpha_s}{pi}G^2\right\rangle = (0012 \pm 0.004)\,GeV^4$ at the renormalization scale $\mu = 1\,GeV$.

Using the mass of the $\rho(3S) = 1.9\,GeV$ [16], we get $\sqrt{s_0'} = 1.9\,GeV$, but in this case, the decay constant of the excited state is bigger than the decay constant of the ground state, that way the sum rules fails. Furthermore, the maximum value of the $\sqrt{s_0'}$ parameter is 1.66 GeV. That way, the excited state decay constant is a bit lower than the ground state decay constant. The minimum value for $\sqrt{s_0'}$ is 1.56 GeV, because $\sqrt{s_0'} - m_{\rho(2S)}$ reaches the value of 100 MeV.

In this way, we can find the Borel window where the QCDSR is valid. In this case, the Borel window is shown in **Figure 2** and it is calculated by the ratio between Eqs. (17) and (18) for a given $\sqrt{s_0'}$ value and considering the $\rho(1S, 2S)$ masses given by [16]. We can see that to have a good accuracy on our results we have to evaluate the QCDSR in a range of $0.8 \leq \mathrm{M} \leq 2.3$, where the pole contribution is bigger than 40%.

In **Figure 3** we display the masses of the $\rho(1S, 2S)$ mesons as function of the Borel mass for three different values of the $\sqrt{s_0'}$ parameter that are: 1.66 GeV (polygonal blue point), 1.61 GeV (red dot-dashed line), 1.56 GeV (diagonal green cross point) and the grey lines representing the masses of the ground state and the first excited state for the ρ meson according to [16]. We can see that for the first

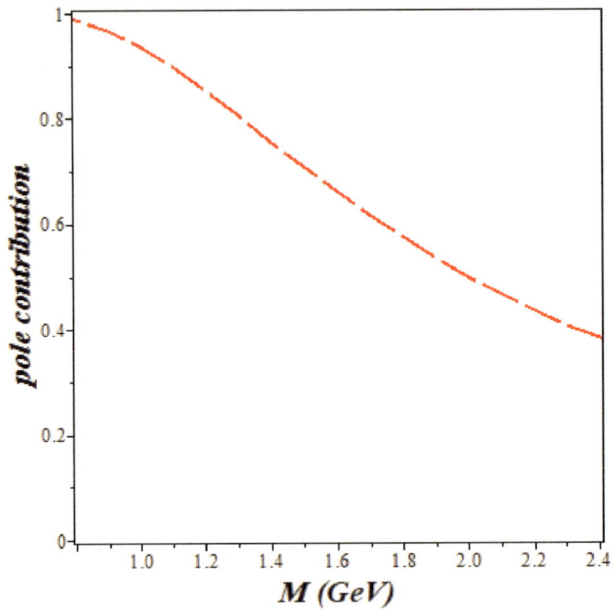

Figure 2.
The red dashed line represents the pole contribution as function of the Borel mass. Note that, for $\sqrt{s'_0} =$ **1.66 GeV** *and the* $\rho(1S, 2s)$ *meson masses given by [16], the range where the QCDSR is* 8GeV ≤ M ≤ 2.3GeV, *where the polo contribution is bigger than 40%.*

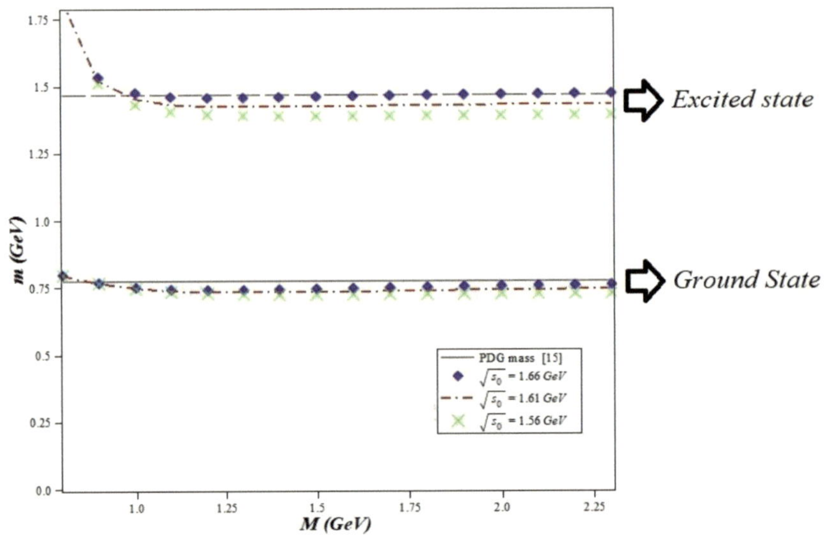

Figure 3.
The masses of the $\rho(1S, 2S)$ *mesons as function of the Borel mass for three different values of the* $\sqrt{s'_0}$ *parameter where* 1.66 GeV *is given by polygonal blue point,* 1.61 GeV *by red dot-dashed line,* 1.56 GeV *by diagonal green cross point and the grey lines representing the masses of the ground state and the first excited state for the* ρ *meson according to [16].*

excited state the mass average is closely to the experimental data for the $\rho(2S)$, where its mass is about 1.45 GeV [16].

For the ground state, in **Figure 3**, we show that the mass average of the $\rho(1S)$ is about 750 MeV, also pretty close to that one seen in [16], that is about 775 MeV for the experimental data.

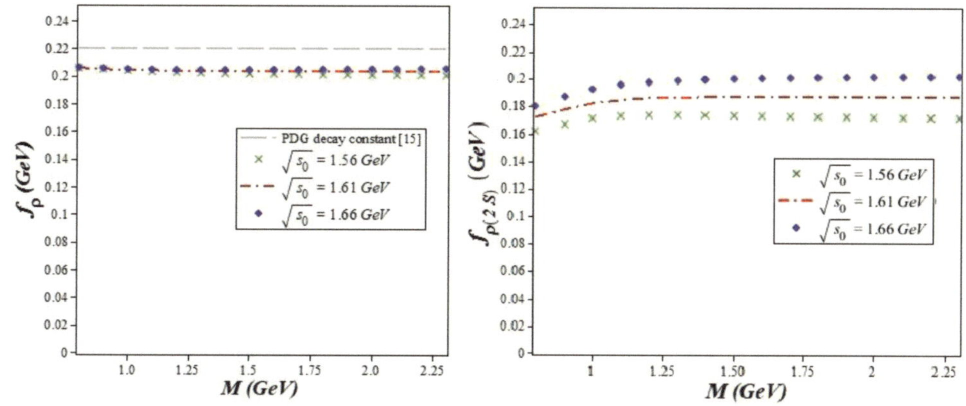

Figure 4.
On the left side, we have the value of the decay constant for the $\rho(1S)$ *meson about* $(203 + 5)$ **MeV**. *The values of the* $\sqrt{s'_0}$ *parameter are: 1.66* GeV *(polygonal blue dot line), 1.61* GeV *(red dash-dotted line) and 1.56* GeV *(diagonal green cross dot line), the grey dashed line is the experimental value [16] that is 220* MeV. *On the right side, we have the value of the decay constant for the* $\rho(2S)$ *meson about* $(186 + 14)$ **MeV** *for the same values of the* $\sqrt{s'_0}$ *parameter are: 1.66* GeV *(polygonal blue dot line), 1.61* GeV *(red dash-dotted line) and 1.56* GeV *(diagonal green cross dot line).*

To evaluate the decay constants, we use the experimental masses given by [16]. For the $\rho(1S)$ we use $0.775\ GeV$ and for the $\rho(2S)$ we use $1.46\ GeV$. In **Figure 4**, we display the decay constant for the $\rho(1S)$ and $\rho(2S)$ mesons as function of the Borel mass for three different values of the $\sqrt{s'_0}$ parameter.

In **Figure 4**, on the left side, we have an average for the decay constant of the $\rho(1S)$ about $(203 \pm 5)\ MeV$, note that the maximum value for the decay constant is that one where $\sqrt{s'_0} = 1.66\ GeV$ (polygonal blue point), the grey dashed line represents the experimental data [16] for the $\rho(1S)$ decay constant that is $220\ MeV$. On the right side, we have an average for the decay constant of the $\rho(2S)$ about $(186 + 14)\ MeV$, where we considered uncertainty with respect to $\sqrt{s'_0}$ parameter at $M = 2\ GeV$.

Furthermore, it is interesting note that in Ref. [20] we can see another way to extract the experimental decay constant of the $\rho^{\pm}$ from semileptonic decay, $\tau^{\pm} \to \rho^{\pm}\nu_\tau$. Note that in Ref. [19].

6. Conclusions

In this work, we made a little revision about the QCD Sum rules method and presented a new method for calculation of the hadronic parameters like mass and decay constant [19] as function of the Borel mass.

We show that the double pole method on QCDSR consists in a fit on the interpretation of the correlation function by the phenomenological side, where the relations dispersion is now presented with two poles plus a continuum of excited states, being these two poles representing the ground state and the first excited state.

For the ρ(1S, 2S) mesons we had a good approximation for the calculations of these masses comparing with the experimental data on the literature. Beyond that, for the decay constant of the *ρ(2S)* meson we had a good prediction like it is seen in [19] where $f_{\rho(2S)} = (182 \pm 10)$ MeV.

Our intention with this work consists on the studying of the vector mesons testing the accuracy of the double pole method and apply this method to analyze others kind of mesons such as scalar mesons.

Acknowledgements

We would like to thank Colégio Militar de Fortaleza and the Universidade Federal de Campina Grande by technical and logistical support.

Conflict of interest

The authors declare no conflict of interest.

Author details

Mikael Souto Maior de Sousa[1*] and Rômulo Rodrigues da Silva[2]

1 Colégio Militar de Fortaleza – CMF, Fortaleza, Brazil

2 Universidade Federal de Campina Grande – UFCG, Campina Grande, Brazil

*Address all correspondence to: mikael.souto@ufrr.br

References

[1] V.A. Novikov, L.B. Okun, M.A. Shifman, A.I. Vainshtein, M.B., Voloshin and V.I. Zakharov, Phys. Rept. 41, 1, 1978.

[2] V.A. Novikov, L.B. Okun, M.A. Shifman, A.I. Vainshtein, M.B., Voloshin and V.I. Zakharov, Phys. Lett. B 67, 409, 1977.

[3] M.A. Shifman, A.I. Vainshtein and V. I. Zakharov, Nucl. Phys. B 147, 385, 1979.

[4] M.A. Shifman, A.I. Vainshtein, M.B. Voloshin and V.I. Zakharov, Phys. Lett. B 77, 80, 1978.

[5] P. Colangelo and A. Khodjamirian, in *At the frontier of particle physics*, ed. by M. Shifman. arXiv:hep-ph/0010175, Vol. 3, pp. 1495-1576.

[6] L.J. Reinders, H. Rubinstein and S. Yazaki, Phys. Rept. 127, 1,1985.

[7] J.P. Singh, F.X. Lee, Phys. Rev. C 76, 065210, 2007. arXiv:nucl-th/0612059

[8] P. Gubler and M. Oka, Prog. Theor. Phys. 124, 995, 2010. arXiv:1005.2459 [hep-ph]

[9] D. Harnett, R.T. Kleiv, K. Moats and T.G. Steele, Nucl. Phys. A 850, 110, 2011. arXiv:0804.2195 [hep-ph]

[10] A.P. Bakulev and S.V. Mikhailov, Phys. Lett. B 436, 35, 1998. arXiv:hep-ph/9803298

[11] A.V. Pimikov, S.V. Mikhailov and N. G. Stefanis, arXiv:1312.2776 [hep-ph]

[12] K. Ohtani, P. Gubler and M. Oka, AIP Conf. Proc. **1343**, 343, 2011. arXiv: 1104.5577 [hep-ph]

[13] P. Gubler, K. Morita and M. Oka, Phys. Rev. Lett. 107, 092003, 2011. arXiv:1104.4436 [hep-ph]

[14] K. Suzuki, P. Gubler, K. Morita and M. Oka, arXiv:1204.1173 [hepph]

[15] W. Greiner, S. Schramm and E. Stein. Quantum Cromodynamics. Ed Springer, 2002.

[16] Particle Data Group: J. Beringer et al., Phys. Rev. D, 86, 010001, 2012.

[17] S. Godfrey, N. Isgur, Phys. Rev. D, 32, 189,1985.

[18] D. Ebert, R.N. Faustov, V.O. Galkin, Phys. Rev. D, 79, 114029, 2009. arXiv: 0903.5183 [hep-ph]

[19] M. S. Maior de Sousa and R. Rodrigues da Silva. Braz. J. Phys, 46, 730-739, 2016. DOI: 10.1007/s13538-016-0449-9

[20] D. Becirevic, V. Lubicz, F. Mescia and C. Tarantino, JHEP 0305, 007, 2003. arXiv:hep-lat/0301020

Chapter 6

Application of Einstein's Methods in a Quantum Theory of Radiation

Richard Joseph Oldani

Abstract

Einstein showed in his seminal paper on radiation that molecules with a quantum-theoretical distribution of states in thermal equilibrium are in dynamical equilibrium with the Planck radiation. The method he used assigns coordinates fixed with respect to molecules to derive the A and B coefficients, and fixed relative to laboratory coordinates to specify their thermal motion. The resulting dynamical equilibrium between quantum mechanical and classically defined statistics is critically dependent upon considerations of momentum exchange. When Einstein's methods relating classical and quantum mechanical statistical laws are applied to the level of the single quantum oscillator they show that matrix mechanics describes the external appearances of an atom as determined by photon-electron interactions in laboratory coordinates, and wave mechanics describes an atom's internal structure according to the Schrödinger wave equation. Non-commutation is due to the irreversibility of momentum exchange when transforming between atomic and laboratory coordinates. This allows the "rotation" of the wave function to be interpreted as the changing phase of an electromagnetic wave. In order to describe the momentum exchange of a quantum oscillator the Hamiltonian model of atomic structure is replaced by a Lagrangian model that is formulated with equal contributions from electron, photon, and nucleus. The fields of the particles superpose linearly, but otherwise their physical integrity is maintained throughout. The failure of past and present theoretical models to include momentum is attributed to the overwhelming requirement of human visual systems for an explicit stimulus.

Keywords: Einstein's quantum theory, matrix mechanics, wave mechanics, momentum exchange, conservation laws, non-commutation, wave function, Schrödinger wave equation, Lagrangian

1. Introduction

Two possibilities are available in the literature for describing the interaction of matter and radiation, classical theory and nonrelativistic quantum theory. Classical theory explains the continuous aspects of electromagnetic radiation, Maxwell's laws, and the theory of heat. Quantum theory explains the Planck radiation law of black body radiation, the discrete nature of observables, and the statistical properties of matter. A third possibility that has remained relatively obscure as an alternative derives from Einstein's 1917 paper "Quantum theory of radiation" and includes aspects of both theories [1]. He shows there that as a consequence of the conservation of momentum the velocity distribution of molecules emitting black body radiation *quantum mechanically* are in dynamical equilibrium with the *classically* derived Maxwell-Boltzman velocity distribution due to thermal exchange. The internal energy distribution of the molecules demanded by quantum theory is then in strict conformance with the emission and absorption of radiation. Because the link between quantum mechanical and classical properties of matter is statistically defined and applies to material systems rather than individual atoms Einstein's theory is considered to be unfinished. A description of atomic structure using his methods is sought after here as a way to fulfill these ideas.

The authors of nonrelativistic quantum theory adopted Einstein's ideas for the A and B coefficients, which are determined by the classical field effect resonance, and described the discrete transfer of energy from a radiation beam to an atomic state; but they neglected the effect of momentum exchange required by the conservation of momentum. The momentum of a photon E/c causes an atom or molecule to recoil in the direction of the beam when it is absorbed and in the opposite direction when it is emitted. Nonrelativistic quantum mechanics places primary importance on the observable properties of radiation in the form of energy measurements while ignoring the more subtle effects of momentum which are more difficult to observe. Consequently the Schrödinger wave equation is formulated continuously without provision for transmitting the discontinuous impulses of photons. The relationship between classical and quantum mechanical statistics that Einstein had carefully constructed breaks down so that instead of a gradual evolution of ideas in which classical and quantum concepts develop together a complete break from classical theory occurred. In the absence of an underlying classical foundation different interpretations of quantum mechanics developed which use methods drawn from facts that are supported by experiment specific to that model alone, but show no relationship to each other. No model has emerged that can account for all the facts. In the following discussion we shall see that the reason no single model of quantum mechanics is able to explain all of the experimental facts, discrete and continuous, yet they concern the same topic is that each one addresses a different aspect of the *same* physical phenomenon; the interaction between matter and radiation.

2. Matrix mechanics

2.1 Historical perspectives

The Bohr model of the atom gives the quantum rule for changes in energy state $E_2 - E_1 = h\nu$, but says nothing about the processes of emission and absorption. Improved understanding of radiation came gradually as experimental techniques improved. Einstein's 1917 paper marks the beginning of quantum mechanics since all subsequent research on the absorption, emission, and dispersion of radiation is based upon it [2]. Through the use of thought experiments and results obtained in an earlier paper on Brownian motion he showed how the microscopic structure of matter is able to influence matter macroscopically. The induced absorption of black body radiation occurs continuously due to random inputs of momentum from thermal collisions and radiation, while induced and spontaneous emission occurs discretely according to the Bohr frequency rule for changes in state and is directed along an infinitesimal solid angle consistent with the photon's recoil momentum E/c. A dynamic equilibrium is thereby created between the thermal energy absorbed by molecules and the quantum mechanical emission of radiation.

Although the A and B coefficients of Einstein's radiation theory have been incorporated into nonrelativistic quantum mechanics the transition of energy from a classical thermal origin to the discrete energy states of atoms and molecules creates discontinuities that are not accounted for by the Schrödinger wave equation. A vastly improved knowledge of the mechanical properties of photons due to momentum in the astronomical sciences, molecular manipulation, optical tweezers, and laser cooling; technological advances has not translated into an understanding of how to incorporate momentum into the equations of quantum mechanics. The momentum of light is treated separately from energy, and Einstein's theory of radiation is the only one that makes explicit use of it when describing absorption and emission. To see why this is true it will be necessary to examine the historical origins of quantum mechanics.

The dynamic equilibrium between classical and quantum mechanical statistical laws that exists for black body radiation is closely related to the phenomenon of dispersion. Dispersion is the continuous change in the angle of refraction of different frequencies of light by a prism or other medium. Although light disperses continuously across the entire spectrum, at certain specific frequencies characteristic of the medium, it is completely absorbed forming lines. When Bohr introduced his theory of electron orbitals he immediately recognized the possibility that the discrete lines of atomic spectra are related to the discrete lines in dispersion phenomena [3]. Other researchers, in particular Debye and Sommerfeld, were also inspired by that possibility and a series of papers appeared that tried to explain the discrete and continuous properties of dispersion by introducing classically inspired modifications of the electron orbitals [2, 4, 5]. However, when experiments revealed that the characteristic frequencies of anomalous dispersion coincide with the frequencies of the spectral lines it was evident that orbiting electrons could not account for both and a complete break from classical theory was necessary. Ladenburg was the first to suggest how the new quantum theory would appear by following Einstein's reasoning leading to the A and B coefficients [5–7]. This enabled him to equate two theoretical expressions, the energy absorbed/emitted by N classical resonators and the energy absorbed/emitted by N′ quantum atoms. By obtaining a statistical balance between classical and quantum mechanical energy exchange he satisfied the conservation of energy, but not that of momentum. Four years later Kramers reinterpreted Ladenburg's results by using the Bohr model of the atom as a multiply periodic system of virtual oscillators [2, 5, 8, 9]. In that model a quantum mechanical variable X is described with a classical Fourier series, where A(n, n- τ) is the quantum analog of the classical amplitude, n indicates the electron orbital number, and τ assumes integral values to denote positive or negative transitions [9].

$$X = \sum_{\tau} A(n, n - \tau) \exp\left[2i\pi\nu(n, n - \tau)t\right], \tau = \mp 1, \mp 2, \dots \tag{1}$$

The Bohr-Kramer method distanced itself from that of Einstein in an important way. Einstein argued that momentum conservation is what distinguishes classical properties observed in laboratory coordinates from quantum mechanical properties observed in atomic coordinates. The discrete and continuous properties of matter are thereby separated from each other physically. In the interpretation by (1), on the other hand, matter-radiation interactions are described exclusively in laboratory coordinates. Fields are described classically by means of Fourier series while quantization is imposed on the field energy. Quantization is thereby understood to be a localization of energy even though the fields extend to infinity and are therefore diffuse. The concept of photon momentum, a property whose displacement in time is *directional*, is replaced by a wave model that is *isotropic* and treats emission as a spherically symmetric process with no net momentum transfer, and processes that are *reversible* in time.

Once Kramer had reinterpreted Einstein's quantum theory of radiation with fictitious harmonic oscillators Heisenberg was able to use it to formulate a theory of quantum mechanics that reconciles the continuity of radiation fields with the discrete energy states of an atom [2, 7, 9]. The complex sets of mathematical rules that he used to describe the frequencies and intensities of spectral lines, may be expressed in the form of a matrix.

$$\sum_k (p_{nk}q_{km} - q_{nk}p_{km}) = \begin{matrix} i\hbar \, for \; n = m \\ 0 \, for \; n \neq m \end{matrix} \tag{2}$$

Each matrix element represents a pair of energy states of the type (1) with the observable properties, frequency and intensity, of an electromagnetic wave. The complete matrix has an infinite number of components and corresponds in its entirety to one of the dynamic variables; the coordinates, the momenta, or the velocities of the particles. The matrix products do not commute as they do in classical theory. When $n = m$ the elements are diagonal and the value of the equation is equal to iħ. For non-diagonal elements, $n \neq m$, and its value is zero.

The quantum mechanical reformulation of the classical Fourier series (1) and (2) is further simplified into its more familiar form by replacing the summed elements with single terms.

$$\mathbf{pq} - \mathbf{qp} = i\hbar \tag{3}$$

The momentum **p** and position **q** are not numbers; but rather arrays of quantities, or matrices. Each component of the matrix is a Fourier series associated with any two of an infinite number of orbits. Because the orbits may extend to infinity both in space and in time exchanges of momentum are delocalized.

2.2 Classical interpretation of matrix mechanics

After three successive modifications from Ladenburger to Kramers to Heisenberg, Einstein's theory is scarcely recognizable. Mathematical modifications that dilute its physical content are given by the Eqs. (1)–(3). Very little remains of Einstein's carefully crafted relationship between classical and quantum mechanical variables despite the fact that all three reinterpretations and the Eqs. (1) through (3) claim to describe the same physical phenomenon, the interaction between matter and radiation. The theories differ dramatically because the directional properties of emitted radiation due to recoil momentum have been replaced by virtual harmonic oscillators which emit energy isotropically as spherical waves and are reversible in time. The balance between thermal energy and radiative energy maintained by momentum exchange depends on oscillators that absorb thermal energy classically and emit energy quantum mechanically directed along an infinitesimal solid angle with momentum E/c. Virtual oscillators that emit isotropically disrupt the delicate balance between classical and quantum mechanical statistical principles which Einstein had so carefully constructed.

The advantage of using energy rather than momentum in a theory of radiation is its ease of use. Energy is defined as a magnitude, which is easier to describe mathematically, to measure, and to calculate. The advantage of momentum, on the other hand, is that its description provides a more accurate picture of a system's time evolution. Position coordinates are assigned to particles relative to a system of reference in order to specify the direction and magnitude of momentum. The conservation of momentum may then be applied and used to interpret observable phenomena. The Ptolemaic planetary system, for example, introduced fictitious epicycles in violation of the conservation of momentum, but continued to be used for a thousand years because it successfully reproduced what was observed. If astronomers had understood the universal properties of momentum they would have immediately rejected a theory that suggests massive objects could reverse motion in empty space.

Einstein used atomic coordinates fixed with respect to a molecule to derive his A and B coefficients describing momentum exchange during the absorption and emission of energy. The linear momentum of molecules due to thermal impulses is described by introducing a second coordinate system defined with respect to the black body container, that is, in laboratory coordinates. The momentum exchange between the opposing external and internal forces of molecules creates a dynamic equilibrium and allows a clear separation between classical and quantum observables respectively. In contrast, the Bohr-Kramers method describes all observables, discrete and continuous, externally with respect to laboratory coordinates. From Heisenberg's perspective there was no need to treat the discrete spectral lines due to atomic orbitals and the continuous observables due to dispersion phenomena differently, concluding that [10], "Quantum mechanics [is] founded exclusively upon relationships between quantities which are in principle observable."

Dispersion phenomena are observed and measured in laboratory coordinates, and not in the coordinates of an atom. They are given by off-diagonal elements of matrices $n \neq m$ where elements above the diagonal refer to changes in frequency due to energy absorption and elements below the diagonal refer to frequency changes due to energy emission. The elements represent the continuously variable resonances of radiation with an atom's valence electrons. The energy of an absorption offsets the energy of an emission except for a difference in phase so a value of zero is obtained for Eq. (2). On the other hand, the diagonal elements of matrices for $n = m$ are real eigenvalues representing ground state energy levels. Absorption results in stimulation to a higher orbital and the subsequent emission of a photon upon decay according to the Bohr frequency condition. The off-diagonal interactions due to continuous momentum exchanges are governed by the Compton equation $p\lambda = h$. Each matrix element is a photon-electron interaction obtained by resolving the Fourier series (1) into its individual components. It is hypothesized that the complete matrix array expresses the conservation of momentum. Heisenberg mistakenly believed that matrices describe atomic structure, but as Einstein showed atomic structure must be described by internally defined coordinates in the unobservable space–time of an atom. To compare atomic and laboratory coordinates a transformation of coordinates must be performed. Transformations may be visualized with the assistance of the electron oscillator shown in the figure.

2.3 Non-commutation

To see how non-diagonal and diagonal elements differ we introduce the idea of an electron oscillator in the figure below. If an electron is raised from the ground state $|1\rangle$ to an excited state $|2\rangle$ and then returns a photon is irreversibly emitted. This is shown schematically in the figure below, where 1 and 2 denote the states and arrows refer to transitions. On the left the energy of an electron increases and then decreases, while on the right the reverse occurs. Each arrow represents one-half cycle of the electron oscillator. If the arrows are used to describe off-diagonal matrix elements, they refer to different atoms. If the elements are diagonal they refer to the same atom. It is a simple way of comparing the laboratory coordinates of matrix elements, as determined by photons, to coordinates of atomic structure determined by electron shells during the absorption and emission of radiation. Although the final state of the quantum system differs the two processes are identical when described in terms of energy differences.

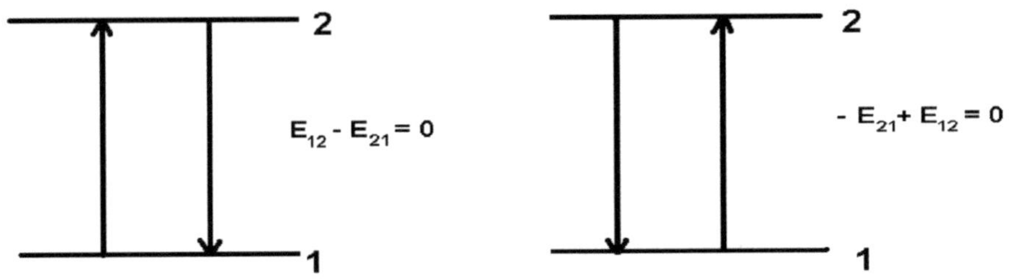

Now consider what happens when the same two energy exchanges are analyzed in terms of the momentum. Using Compton's equation for the momentum of a photon, $p = h/\lambda$, the first exchange may be expressed:

$$p_{12}\lambda_{12} - p_{21}\lambda_{21} = 0 \tag{4}$$

Angular momentum increases by an amount $\hbar$ when the electron is excited and is then reduced by the same amount when the atom returns to its ground state $|1\rangle$. Thus this type of photon emission ends up with the atomic system in its ground state.

However, when the order of the electron transitions is reversed on the right of the figure we see by the following expression that a description of momentum exchange gives a different result.

$$p_{21}\lambda_{21} - p_{12}\lambda_{12} = \hbar \tag{5}$$

The electron begins in an excited state $|2\rangle$, reverts to the ground state $|1\rangle$ by emitting a photon, and is excited once again. Thus the final state of the atomic system has an angular momentum that is greater than the ground state by an amount $\hbar$. In both cases (4) and (5) a photon is emitted, but because the order of the physical variables changed the angular momentum of the atomic system described by (5) is greater than (4) and the physical variables do not commute. Non-commutation is interpreted as the irreversibility of momentum when transforming between atomic and laboratory coordinates.

3. Wave mechanics

3.1 Historical perspectives

Einstein introduced the founding principles of wave mechanics with concepts from his 1905 papers on special relativity and the photoelectric effect which de Broglie extended to material particles. He also provided the stimulus which led to completion of these ideas in a series of papers on the quantum theory of gases by showing that the same statistics Bose had applied to light quanta could also be used to describe emission from a monatomic ideal gas [11]. This led directly to the further development of wave mechanics by Schrödinger and the introduction of the wave function who openly acknowledged his indebtedness to Einstein in a letter [12]. "By the way, the whole thing would not have started at present or at any other time (I mean as far as I am concerned) had not your second paper on the degenerate gas directed my attention to the importance of de Broglie's ideas." His papers also stimulated Dirac to write the first paper on quantum electrodynamics introducing the concept of second quantization [13]. Despite the implicit dependence of their theories upon his ideas none of them heeded his advice about momentum [1]. "Most important appears to me the result about the momentum transferred to the molecule by incoming and outgoing radiation." If they had followed Einstein's logic a more coherent description of quantum mechanics would have emerged.

3.2 Physical interpretation of the wave function

The concept of electron oscillator may be used to describe the rotation of the wave function of half-integer spin particles [14]. Excitation consists of the rotation of an electron's wave function through 2π radians during the *absorption* of one complete cycle of an electromagnetic wave. Decay corresponds to a second rotation of 2π radians during the *emission* of a complete wave cycle. In other words, a complete electron cycle, excitation and decay, consists of two wave function rotations, or 4π radians, and two cycles of an electromagnetic wave, where rotation refers to a change in phase of the electromagnetic field rather than a change in physical space. The electron begins its cycle during energy absorption by entering into a superposition state with a photon's sinusoidal electromagnetic fields and it exits the superposition state when the photon is released. The completed rotation consists of one cycle of an electron oscillator and two cycles of a wave. Thus changes in state can be viewed variously as the excitation and decay of an electron, photon creation and annihilation, superposition of fields, or cycling of a wave; depending upon which physical aspect of the phenomenon one chooses to describe. We use imaginary numbers to describe the transformation of coordinates from the atom to ordinary space so that it is possible to describe the rotation of a wave function mathematically.

The transfer of photon momenta to molecules in induced absorption and emission was predicted theoretically by Einstein and has been verified macroscopically by experiments of many types. It has also been verified microscopically by recent experiments with ultracold three-level artificial atoms which support the idea that momentum is a necessary parameter for the description of emission processes [15]. In the quantum Zeno effect frequent measurements arrest the progress of a "quantum jump". The measurements are equivalent to momentum exchange thereby confirming the earlier hypothesis that photon momenta need to be included in theories of the stimulated absorption and emission of radiation. An incoming photon transfers a momentum $+E/c$ to an atom in the ground state and superposes its fields with an electron's fields. When it exits the superposition state it transfers recoil momentum $-E/c$ to the atom and is expelled. The induced absorption and emission momenta are applied at different locations, the ground state electron shell and the excited state electron shell; and they are directed in opposite directions. Taking momentum into account during the time evolution of absorption and emission processes suggests that the electron oscillator cycles at discrete points in space due to momentum exchange and discontinuously in time.

In the wave mechanical view emission occurs by discrete energy exchange, but momentum exchange is either undetectable or does not occur; a situation that is refuted by the Einstein theory of radiation and cannot be sustained by experiment. The Schrödinger wave equation must be reformulated to reflect the discontinuous spatial coordinates and asymmetry of time necessary for momentum exchange.

3.3 Lagrangian model

The matrix mechanical observables of matter-radiation interactions are described in laboratory coordinates, while wave mechanical properties of matter are described in atomic coordinates. Both describe the same characteristic, the steady states of an atom, but they approach them from different points of view; external and internal. The Einstein theoretical model of matter-radiation interactions adopts both points of view simultaneously within a single material system, as the dynamic equilibrium between external and internal forces. To describe the radiative processes of a single atom a wave equation is needed that includes photons, describes the time evolution of the wave function, and explains how discrete exchanges of momentum can occur during stimulated absorption and emission. Finally, in order to be in agreement with special relativity theory it must be symmetric in the space and time coordinates.

It is possible to formulate a relativistic wave equation by taking the action integral of a Lagrangian $S = \int L dt$. Dirac has previously advised on the proper use of the Lagrangian in quantum mechanics [16], "we ought to consider the classical Lagrangian not as a function of the coordinates and velocities but rather as a function of the coordinates at time t and the coordinates at time t + dt." Following Dirac's initiative we let the coordinates at time t and at time t + dt denote electron shells corresponding to the states $|1\rangle$ and $|2\rangle$ respectively. Next, "We introduce at each point of space-time a Lagrangian density, which must be a function of the coordinates and their first derivatives with respect to x,y,z, and t corresponding to the Lagrangian in particle theory being a function of coordinates and velocities. The integral of the Lagrangian density over any (four-dimensional) region of space-time must then be stationary for all small variations of the coordinates inside the region, provided the coordinates on the boundary remain invariant"; where the "four-dimensional region of space-time" refers to the area between electron shells and "the coordinates on the boundary" refers to the electron shells. Absorption initiates from the steady state $|1\rangle$ with coordinates $r_1 = (x_1,y_1,z_1)$ and time t_1, and it finalizes at $|2\rangle$ with coordinates $r_2 = (x_2,y_2,z_2)$ and time t_2; where r_1 and r_2 denote electron shells. The Lagrangian density within the four-dimensional space–time region bounded by the electron shells is a function of the coordinates and their first derivatives $L(\phi_i, \phi_{i,\mu})$. The conditions are satisfied by an action integral of the Lagrangian density.

$$S[\phi_i(t)] = \int_{r_1}^{r_2}\int_{t_1}^{t_2} L\left(\phi_i, \phi_{i,\mu}\right) d^3x dt = h \tag{6}$$

The action is a functional, a function of the values of coordinates on the *discrete* boundaries of the space–time surfaces r_1 and r_2 which are in turn functions of the *continuous* space–time variables of the fields within the surface. The discrete space–time variables assigned to the limits of integration describe electron shells and the continuous space–time variables of the Lagrangian density describe electromagnetic fields. Thus the photon is represented as a four-dimensional localization of field within the electron shells. Momentum exchange occurs when a photon makes contact with a point on the electron shell whether by absorption or emission· Even though complementarity denies the simultaneous presence of wave and particle properties in free space, they are present in atomic space when a photon's sinusoidal fields are localized within electron shells.

3.4 Physical model of the atom

If the photon is created as an independent entity when energy is absorbed; then quantum mechanics refers to not two, but three bodies. It presumes that the three field sources are loosely bound within a conservative, or frictionless system, that they are free to interact with each other, and that each of the three particles contributes to the atomic system independently. For the related case of three particles with gravitational fields no general closed form solution is possible [17]. Gravitationally bound three-body systems result in chaos for nearly all initial conditions. It should not be surprising therefore that a physical system consisting of three electromagnetic field sources; electron, photon, and nucleus; also has an indeterminate outcome. To obtain the equations of motion for an electromagnetic three-body problem when the only knowledge available about the particles is their field properties, we need to obtain a series of partial solutions, which are the different mathematical models. Because an exact solution is not possible for the dynamic evolution of a three body system all solutions are considered approximations.

The three-body model of atomic structure may be described formally by introducing a wave-like, physically independent field source ε, the localized photon, into our description of excited atomic states. The modified Hamiltonian is now given by,

$$H = T + \varepsilon + V \tag{7}$$

where T refers to an electron, ε represents a "captured" photon, and V represents the nucleus. Each of the three field sources (or particles), possesses a unique vector field; that is, a well-defined field geometry, while the plus and minus signs indicate that the superposition of fields is linear. The Eq. (7) contains the essence of quantum mechanics as a three-body conservative system in real space, as opposed to nonrelativistic descriptions in abstract space. The equations revert to their classical two-body form when the influence of ε is negligible.

4. Conclusion

If momentum is not taken into account the structure of an atom and its observable properties may be described in the same space. In other words, we can plot the motion of a hydrogen atom's electron in the same space as the motion of the nucleus. If momentum is included a single space–time no longer suffices. When a photon interacts with an atom its linear momentum is transformed into angular momentum and an electron is excited. The angular momentum can no longer be described in laboratory coordinates and instead is expressed in atomic coordinates. All matter has internal and external aspects that are described in distinct coordinate systems. The idea of internal and external properties of matter is as old as science itself having first been expressed by Socrates and Aristotle; however, by introducing Eq. (6) it is proposed as a universal property of matter. Only Einstein fully grasped the need for distinct coordinate systems to describe matter through his theories of the photoelectric effect and Brownian motion. He concluded his quantum theory of radiation by stating [1], "For a *theoretical* discussion such small effects [due to momentum] should be considered on a completely equal footing with the more conspicuous effects of a radiative *energy* transfer, since energy and momentum are linked in the closest possible way." His advice was not fully appreciated due to an inability to visualize the time evolution of a radiating atom.

The conscious mind requires mental images to be able to understand and describe natural phenomena. "For Plato says that we would be engaging in futile labor if we tried to explain these phenomena without images that speak to the eyes." [18]. The need for visual images forms the foundation of classical physics and is the source and origin of science itself. All stages of formulating a theory; whether observation, analysis, or experiment; is intimately connected to the visual system. In fact the visual cortex is so dominant an area of the brain that when blindness occurs it processes tactile and auditory sensory data instead. Visualization was an important factor during the derivation of quantum mechanics and as well of scientific theory in the past. The need to visualize explains why Heisenberg insisted on a theory of "observables", and it also explains why wave mechanics quickly became more popular. It also accounts for the fact that none of the mathematical models explicitly includes the photon.

Author details

Richard Joseph Oldani
Clymer, NY, USA

*Address all correspondence to: oldani@juno.com

References

[1] Einstein, A. Phys. Z. **18**, 121 (1917). https://einsteinpapers.press.princeton.edu/vol6-trans/232

[2] van der Waerden B. L. *Sources of Quantum Mechanics* North -Holland, Amsterdam, 1967), p. 4.

[3] Bohr, N. On the constitution of atoms and molecules Phil Mag **26** (151), 1–24. DOI:10.1080/14786441308634955

[4] Sommerfeld, A. "Quantum theory of spectral lines" Ann Phys **51**, 125. (1916)

[5] Taltavullo Jordi, M. "Rudolf Ladenburg and the first quantum interpretation of optical dispersion" *Eur Phys J.* H **45**, 123 (2020).

[6] Ladenburg, R.. "The quantum theoretical meaning of the number of dispersions electrons" Z. Phys. **4**: 451–468 (1921).

[7] Jammer, M. The Conceptual Development of Quantum Mechanics (Tomash, 1989)

[8] Kramers, H. "The law of dispersion and Bohr's theory of spectra" Nature **113**, 573 (1924).

[9] J. Mehra & H. Rechenberg, *The Historical Development of Quantum Theory,* Vol. II (NY: Springer, 1982-1988), p. 172.

[10] Heisenberg W. (1925), *Z Phys* **33**, 879 in B.L. van der Waerden (ed.), *Sources of Quantum Mechanics*, (Dover, 1968), p. 261.

[11] Einstein, A. *Sitz Preus Akad d. Wiss* (1924), p. 261; (1925), p. 3.

[12] Schrödinger, E. (1926) in M. Jammer, *The Conceptual Development of Quantum Mechanics* 2nd ed. (NY: Tomash, 1989), p. 258.

[13] Bromberg, J. in C. Weiner (ed.), *History of Twentieth Century Physics*, (NY: Academic, 1977), p. 147.

[14] Schiff, L.I. *Quantum Mechanics*, (NY: 1968), p.205.

[15] Minev, Z., Mundhada, S., Shankar, S. *et al.* "To catch and reverse a quantum jump mid-flight." Nature **570**, 200 (2019). https://doi.org/10.1038/s41586-019-1287-z

[16] P.A.M. Dirac, Phys Zeit Sow **3**, 1 (1933).

[17] Barrow-Green J. in Gowers, T.; Barrow-Green, J.; Leader, I. (eds.) The Princeton Companion to Mathematics, (Princeton U. Press, 2008), p. 726.

[18] Theon of Smyrna, 1st century